Learning *from* Dogs

Innate Wisdom from Man's Best Friend

Paul Handover

Learning from Dogs
Innate Wisdom from Man's Best Friend

Softcover: #978-0-9967782-0-6
Ebook: #978-0-9967782-1-3

Dedication

This book is dedicated to my
German Shepherd dog, Pharaoh,
and to all the loving dogs that my wife
and I have had the pleasure of knowing,
and also to all the wonderful dogs
in the whole wide world.

Contents

Foreword

Some 13.8 billion years ago out in the vast nothingness of space, there was a big bang, a humongous gargantuan explosion from which all matter in the universe originated. Every element, atom, molecule, subatomic particle in the known universe spewed forth from this fantastic explosion. Eventually some of this stardust coalesced into billions of galaxies containing tens of billions of stars around which revolved hundreds of billions of planets.

On one of these planets, a rather ordinary "pale blue dot" as Carl Sagan put it, in an average unremarkable galaxy in an obscure corner of the universe, something magnificent and extraordinary transpired. It just so happened that this very ordinary orb was in what astronomers describe as the "Goldilocks Zone": where the temperature was not too hot, not too cold. And in the primordial stew in which this planet simmered was just the right mixture of all the ingredients necessary for the creation of self-replicating molecules and, voilà, life emerged.

In the slow but inexorable and unrelenting march of evolution, fine-tuned by genetic mutation and culled by natural selection, these molecules gradually over the next few billion years formed single cells, and then colonies of cells, then multicellular organisms culminating in the higher plants and animals, most notably the self-proclaimed "pinnacle of creation": *HOMO SAPIENS*.

Many species came and many species went, by way of mass extinctions, but life (primitive as it was) plodded along for eons and epochs. And sometime very late in this long and arduous four-billion year evolutionary journey, a wondrous and extraordinary relationship was forged between two distantly related species: humans and dogs. In his book *LEARNING FROM DOGS*, Paul Handover devotes a whole chapter to dog origins, in which he discusses in great detail the differing theories regarding how and when this unique relationship may have developed. Regardless of the details, this was a very special relationship, unlike any interaction between any two species ever in the history of the earth, and the relationship continues to this day, stronger and more poignant than ever.

Initially, the relationship seems to have been mutually beneficial, enhancing the survival and well-being of both species. But now, as Paul alludes to, I think in many ways it is we humans who have benefited most. As any dog lover can attest to, the human-dog bond is a deeply meaningful experience that defies description and can have a profound effect on our hearts and minds.

I am constantly reminded of this when, for example, I have the unenviable duty of having to tell a young family that their 13-year-old Golden Retriever has terminal cancer and they are forced to make the difficult decision to euthanize their dear old companion; to end the life (and suffering) of a loyal loving member of the family. The heartwrenching grief and sense of loss is almost unbearable and testifies to the love and compassion we feel for these animals. Or conversely, when I see the sheer joy, love, and pride in a child's eyes as they present their new puppy to me for its first vaccinations, again I am reminded of the power of this bond.

Suzanne Clothier in her book *Bones Would Rain from the Sky: Deepening Our Relationships with Dogs* writes movingly about this heartfelt connection between dog and human:

> There is a cycle of love and death that shapes the lives of those who choose to travel in the company of animals. It is a cycle unlike any other. To those who have never lived through its turnings or walked its rocky path, our willingness to give our hearts with full knowledge that they will be broken seems incomprehensible. Only we know how small a price we pay for what we receive; our grief, no matter how powerful it may be, is an insufficient measure of the joy we have been given.

But what exactly is this human dog bond and why do we feel such an affinity for this species above all others? My feeling is that it may be associated with our deep but subconscious longing for that age of simple innocence and innate human goodness that we supposedly possessed before we became truly "human"; that child-like innocence or what Jean-Jacques Rousseau referred to as the "noble savage" before being corrupted by civilization, before we were booted out of the Garden of Eden. We humans, for better or worse, somewhere along that evolutionary road acquired consciousness or so-called human nature and with it we lost that innocence.

What we gained were those marvelous qualities that make us uniquely human: a sense of self-awareness; an innate moral and ethical code; the ability to contemplate our own existence and mortality and our place in the universe. We gained the ability to think abstract thoughts and the intellectual power to unravel many of the mysteries of the universe. Because of that acquired consciousness and humans' creative and imaginative mind we have produced the likes of Shakespeare, Mozart, and Einstein. We have peered deep into outer space, deciphered the genetic code, eradicated deadly diseases, probed the bizarre inner world of the atom, and accomplished thousands of other intellectual feats that hitherto would not have been possible without the evolution of our incredible brain and the consciousness with which it is equipped.

No other living species on this planet before or since has developed this massive intellectual power. But this consciousness was attained at what cost? Despite all the amazing accomplishments of the human race, we are the only species that repeatedly commits genocide and wages war against ourselves over political ideology, geographic boundaries, or religious superstition. We are capable of justifying the suffering and death of fellow human beings over rights to a shiny gold metal, or a black oily liquid that powers our cars. We are the only species that has the capability to destroy our own planet, our only home in this vast universe, by either nuclear warfare, or more insidiously by environmental contamination on a global scale.

Was it worth it? No matter what your or my opinion may be, Pandora's Box has been opened and we cannot put the lid back on. But what can we do now to reverse this trend and help improve the quality of life for humanity and ensure the well-being of our planet? In his book, Paul proposes that if we recognize the problem and look very critically at ourselves as a unique species, with awesome powers to do both good and bad, and if we put our collective minds to the task it may be possible to retrieve some of the qualities of that innocence lost without losing all that we have gained.

Dogs represent to me that innocence lost. Their emotions are pure. They live in the present. They do not suffer existential angst over what they are. They do not covet material wealth. They offer us unconditional love and devotion. Although they certainly have not reached the great heights of intellectual achievement of us humans (I know for a fact that this is true after having lived with a Labrador Retriever for several years), at the same time they have not sunk to the depths of depravity to which we are susceptible. It could be argued that I am being overly anthropomorphic, or that dogs are simply mentally incapable of these thoughts. But nevertheless, metaphorically or otherwise, I believe that dogs demonstrate a simple and uncorrupted approach to life from which we all could benefit.

In his book *Learning from Dogs*, Paul repeatedly elaborates on this thought in different contexts: that dogs possess innate qualities of character that it would behoove us to emulate. I think the crux of Paul's thesis is that, within the confines and limitations of our human consciousness, we can (and should) metaphorically view the integrity of the dog as a template for human behavior. He has an epiphany of sorts when his psychotherapist and good friend shares with him the profound observation that dogs are creatures of integrity and it is this that he feels is the central message of the book.

In the following pages Paul Handover has written a sincere and personal account of what he feels we humans might gain if we took the time and effort to reflect on our current human situation and adopted what he sees as the integrity of the dog as an example for our future conduct in order to improve the quality of life on this planet.

Over the years Paul and especially his wife Jean have rescued many, many dogs from Mexican streets that otherwise would have suffered terribly and succumbed to early death. Paul has intimately lived with as many as 14 of these dogs at a time, and with his diverse background of experiences and skills he is uniquely qualified to elucidate and expound on the character of the dog and how it relates to the human condition.

His book is hard to define as a genre. At times his work reads as autobiography, diary, or blog. Other times he reinforces a point with speculative and imaginative fictional narrative. And other times he injects factual research. Occasionally launching into other intellectual tangents, Paul ultimately returns to his central thesis: "What we can (and should) learn from dogs."

I love dogs because they are dogs, not because they are furry little humans who seem to exhibit some of our virtuous personality traits. I love the smell of puppy breath. I love an exuberant tail wag and sloppy drool-laden full-tongue kiss upon returning home. I love watching the uninhibited joy of retrieving a tennis ball or a stick thrown into the river. I love the fact that the simple pleasure of a delicious dog food dinner can be the high point of the day. Or that a drive in the countryside with the windows down and lips and ears flapping in the breeze can make a dog giddy with joy. Dogs are neither saints nor sinners. They are dogs, pure and simple, and I love them for it. Maybe we too could still learn to appreciate the simple pleasures of life. As Paul Handover proposes, *Learning from Dogs* may be the answer.

> The dogs in our lives, the dogs we come to love and who (we fervently believe) love us in return, offer more than fidelity, consolation,

companionship. They offer comedy, irony, wit, a wealth of anecdotes, 'the shaggy dog stories' and 'stupid pet tricks' that are the commonplace pleasures of life. They offer, if we are wise enough or simple enough to take it, a model for what it means to give your heart with little thought of return. Both powerfully imaginary and comfortingly real, dogs act as mirrors for our own beliefs about what would constitute a truly humane society. Perhaps it is not too late for them to teach us some new tricks. (Marjorie Garber, *Dog Love*, Simon and Schuster, 1996).

James R. Goodbrod, Master's Degree,
Doctor of Veterinary Medicine

Preface

This is a journey that would not have begun had I not had my eyes opened to the fact that dogs are creatures of integrity. So thank you, J.[1] You and I had no idea what you started back in 2007.

The other undeniable fact is that my blog, from which this book takes its name, would not have commenced in July 2009 if I had not counted Pharaoh, my German Shepherd dog, as my dearest animal friend since 2003. I owe it all to you, dear Pharaoh. Indeed, it was the many people over those years who commented, remarked and regularly read my blog posts that persuaded me to participate in National Novel Writing Month (NaNoWriMo) in 2013 and 2014. I used the Novembers of each of those years to draft in excess of 110,000 words of non-fiction writing that underpinned the creation of this book.

The inspiration behind the title goes back to 2007 when I was living in the village of Harberton, near the town of Totnes, in the southern part of the county of Devon, southwest England. I had cause to work with J and one particular afternoon we were talking about integrity. J made an observation about Pharaoh, my German Shepherd, that was in the same room, albeit fast asleep, and about dogs in general. His observation was that dogs are creatures of integrity. My curiosity was tweaked by J's revelation and I pondered that someday I might write about this. (Indeed, that same day I registered ownership of the domain name: learningfromdogs.com.) Fast-forward to July 2009 when Pharaoh and I were living with Jean, and her 14 dogs, in San Carlos, Mexico, and I started writing my blog. Five years later, when Jean and I were living in southern Oregon, still with Pharaoh and nine other dogs, saw the conception of this book.

Acknowledgments

It is said that one writes a book for oneself but then edits it for the reader. That is a powerful reminder of the critical importance of editing a draft manuscript. The rewriting involved to turn the draft manuscript into a finished book could not have been accomplished without the professionalism of Joni Wilson of Missouri who, for a very reasonable fee in my opinion, edited the manuscript and cover text beyond anything that I could have done on my own. Joni was recommended to me by Deborah Perdue, who trades as Illumination Graphics here in Oregon. The design of the cover and the interior of this book were very professionally undertaken by Deborah.

Then I must give thanks for the huge support, in both time and energy, of local vet and close friend Jim Goodbrod, who very kindly in addition wrote the foreword to this book.

Speaking of writing, I must compliment the fine people at Literature & Latte Limited, in Truro in Cornwall, England, who are the creators of the Scrivener writing software that I used for this writing project. Frankly, I was blown away by the quality of that software.

Then I must thank those wonderful individuals who generously offered to proofread the draft: Deborah Taylor-French; my sister, Eleanor, and her husband, Warwick; Dordie

Lampier; Ira Wiesenfeld; Larry and Janell Little; Karla Powell; and Trish Isles. Your feedback was invaluable. Then Richard Maugham and Chris Snuggs, both dear friends of mine for many years, who volunteered to read the final edited version before it was passed to Joni Wilson.

Thanks also to the members of organisation AIM (Authors Innovative Marketing) in Grants Pass, Oregon, who not only welcomed me before I was a published author but provided valuable advice to this first-time author who felt lost on more than one occasion.

My final and most important thanks go to my beloved wife, Jean. Over weeks and weeks, as I drafted and re-drafted chapter after chapter, Jean patiently listened to me reading out each new piece, giving me honest and valuable feedback and never hesitating to interrupt what she was doing when, often without notice, I sought her views to some aspect of my writing. I would never have completed this book without Jean's honest and loving devotion to the project.

Introduction

The beloved dog, *Canis lupus familiaris*, has been humankind's most glorious companion for thousands of years. This species reveals that special companionship in numerous ways, especially those dogs that are lucky enough to be living with caring and loving humans. Some might argue, and argue correctly, that this characteristic is not exclusive to dogs. Cats, horses, birds, chickens, goats, pigs, and doubtless other animals, can live happily in a domesticated arrangement with humans.

However, this is a book titled *Learning from Dogs* and is largely about dogs and, more specifically, about what I have learnt from a close domestic relationship with so many dogs during a period of more than ten years.

Despite the many thousands of years that the dog has been associated with humans, the origins of the dog are still the subject of research and far from being clear, in a scientific sense. In Part One, I offer what is known and what is still conjecture, including my fictional account of that first contact between man and wolf, the genetic predecessor of the dog. Part One also includes two autobiographical accounts of a relationship with a German Shepherd, one from my childhood days and another some 47 years later.

Demonstrating that this is a book less wholly about dogs and more about the lessons these wonderful animals offer us, in an emblematic or metaphorical manner, in Part Two, I take a look at where humankind is in this 21st century. On so many fronts there are scary views of the future. It feels as if the certainty of past times has gone, as if many of the trusted models of society are failing. Whether we are talking politics, economics, employment, and the environment, radicalisation of opinions seems more prevalent than a common desire to leave things in better shape than when we found them. Added to that, it's as if my generation (I'm a 1944 baby) is grossly unaware that without radical change in how we care for and treat this planet, we might be leaving a hostile world for our grandchildren.

Naturally, such a broad, general statement cannot be true for all people in all places on planet Earth; there are countless good people engaged in countless good causes. Nevertheless, there is little doubt in my mind that many are deeply worried about the future of the world and looking for ways of ensuring a safe future for all.

In Part Three, I endeavour to put that need for a safe future into context in terms of how long humankind might have to implement the required changes to bring about a secure future. In addition, examining the key, essential changes that have to be embraced; describing the foundations, the anchors so to speak, that need to be put in place to guarantee a long future both for humans and for the entire natural world.

The essence of those changes, the central purpose of this book, follows in Part Four. I am aware that in reviewing the attributes of dogs, and how they serve as emblems for humankind, I run two risks: first, of being overly romantic and, second, of mistakenly seeing those attributes in anthropomorphic ways by imparting humanistic qualities to our dogs. My intention is not to make these mistakes but you, dear reader, will be the final judge of that.

Part Five is perhaps, a tad self-indulgent for it offers a deeply personal look at the way that living with dogs, living "cheek by jowl" with so many dogs for so many years, has left an almost sacred tone about me as I live through the years of this final stage of my life. Then in the conclusion to the book I try to draw together the whole endeavour.

Finally, readers will quickly spot that this book is written in the Queen's English, apart from the words of Dr. Jim's Foreword and other Americans quoted herein. Prior to moving to be with Jean, now my wife, in 2008 and later settling down in southern Oregon, for more than 60 years I had lived in the United Kingdom, being born in Acton, North London. Old habits die hard and I trust that non-English, English speakers will not be too annoyed with me.

Part One:

Looking Backwards

Chapter 1

THE HISTORY OF THE DOG

"On one of these planets, a rather ordinary 'pale blue dot' as Carl Sagan put it, in an average unremarkable galaxy in an obscure corner of the universe, something magnificent and extraordinary transpired." So wrote Jim Goodbrod in his Foreword to this book, referring to the most magnificent and extraordinary creation of life.

Let alone that event being beyond any rational understanding of those of us alive in this 21st century, it is almost inconceivable to go back just a mere smidgen of time, to go back the 200,000 years to when the relationship between that ordinary planet, planet Earth, and humans, as in *H. sapiens*, came to pass. Despite the difficulty of sensing such immense periods of time, there is something extremely beautiful in the knowledge that about 100,000 years ago, namely about half-way along that journey between us and our planet, DNA evidence suggests that the animal we today call the dog evolved as a separate species from the grey wolf.

That's about all that we do know, although that's not to say that there aren't plenty of theories. When we look at some breeds of dogs, let's say the Chihuahua, it beggars belief that the wolf was the ancestor of that dog. Not so hard to believe, mind you, when we look at a breed such as the German Shepherd. Many German Shepherd dogs look like they are first cousin to a wolf.

The Latin "binomial nomenclature" for both the wolf and dog offers clarity irrespective of specific dog breed. I am, of course, referring to *Canis lupus* for the wolf and *Canis lupus familiaris* for the dog. For those, like me, who had to refresh their memory of this naming convention, the first part of the name identifies the genus to which the species belongs and the second part identifies the species within that genus. Thus, humans belong to the genus *Homo* and within this genus to the species *Homo sapiens*.

Thus both the wolf and the dog belong to the same genus. However, when we enquire as to when *lupus familiaris* split away from *lupus* then it all becomes much less clear.

Scientific American magazine, in 2009, quoted in an article[2]: "The going theory is that dogs were domesticated somewhere between 15,000 and 40,000 years ago."

Applying a periodic label to those past times, such as the Mesolithic or Palaeolithic periods, is not helpful, because such names for those archaeological periods vary enormously[3] from region to region. Thus it might be clearer for readers if we stay with the number of years involved.

The end of the last glaciation period, the Ice Age, was about 12,000 years ago. That heralded the start of the period when humankind evolved from a hunter-gatherer existence to that of farming and herding. People discovered how to cultivate crops and began to learn how to domesticate animals and plants[4].

There is a view that around 10,000 years ago, when humans started settling down, there was contact with wolves that led to some wolves living on the fringes of human activities and thence the long evolutionary journey to the dog. But it is an understanding that is not fully shared by all in the field. Indeed, Professor Marc Bekoff[5] in a telephone call with me said that of the two theories of the origin of dogs, either from wolves scavenging from early man, or from an evolutionary split, the evidence was overwhelmingly in support of dogs being the result of an evolutionary split from the wolf.

Mark Derr[6] is an American author and journalist, noted for his books about dogs. He is the author of a number of books including *A Dog's History of America* and *Dog's Best Friend*. In 2006 he wrote an article, "The Wolf Who Stayed", that first appeared in *The Bark* magazine.[7]

When I first read the article I realised that there was much information that I hadn't come across before. I contacted Mark and asked if he might grant me permission to include his article in this chapter. Mark generously offered me permission to quote from his article in part or in full. It is such a comprehensive review of the whole history of the dog, known and speculated, that it is included, in full, as an Appendix to this book.

Nevertheless, some of what Mark has written really should be included in this chapter. For instance, these three paragraphs:

> Dates range from the dog's earliest appearance in the archaeological record around 14,000 years ago to the earliest estimated time for its genetic sidestep from wolves around 135,000 years ago. Did the dog emerge in Central Europe, as the

archaeological record suggests, or in East Asia, where the genetic evidence points? Were they tame wolves whose offspring over time became homebodies, or scavenging wolves whose love of human waste made them increasingly tame and submissive enough to insinuate themselves into human hearts? Or did humans learn to follow, herd and hunt big game from wolves and in so doing, enter into a complex dance of co-evolution?

Despite the adamancy of adherents to specific positions, the data are too incomplete, too subject to wildly different interpretations; some of the theories themselves too vague; and the physical evidence too sparse to say with certainty what happened. Nonetheless, some models, and not necessarily the most popular and current ones, more clearly fit what is known about dogs and wolves and humans than others. It is a field in high flux, due in no small measure to the full sequencing of the dog genome. But were I a bettor, I would wager that the winning view, the more-or-less historically correct one, shows that the dog is the result of the interaction of wolves and ancient humans rather than a self-invention by wolves or a "conquest" by humans.

Our views of the dog are integrally bound to the answers to these questions, and, for better or worse, those views help shape the way we approach our own and other dogs. It

> is difficult, for example, to treat as a valued companion a "social parasite" or, literally, a "shit-eater." To argue that different breeds or types of dogs represent arrested stages of wolf development both physically and behaviorally is not only to confuse, biologically, description with prescription but also to overlook the dog's unique behavioral adaptations to life with humans. Thus, according to some studies, the dog has developed barking, a little-used wolf talent, into a fairly sophisticated form of communication, but a person who finds barking the noise of a neotenic wolf is unlikely to hear what is being conveyed. "The dog is everywhere what society makes him," Charles Dudley Warner wrote in *Harper's New Monthly Magazine* in 1896. His words still hold true.

That the dog is descended from the wolf, or, more precisely, "the result of the interaction of wolves and ancient humans", seems to be a pretty conclusive feature of our mutual evolution and history. As Mark Derr writes, that is about all that *is* agreed among those who have set out to answer these fundamental questions about the origins of the dog. So much about the domestication of the dog is uncertain; as Mark puts it, "specifically, the who, where, when, how and why of domestication."

Two years after Mark Derr's article appeared in *The Bark*, the NBC News website in 2008 carried details[8] of what was thought to be the earliest known dog.

> An international team of scientists has just identified what they believe is the world's first

> known dog, which was a large and toothy canine that lived 31,700 years ago and subsisted on a diet of horse, musk ox and reindeer, according to a new study.
>
> The discovery could push back the date for the earliest dog by 17,700 years, since the second oldest known dog, found in Russia, dates to 14,000 years ago.
>
> Remains for the older prehistoric dog, which were excavated at Goyet Cave in Belgium, suggest to the researchers that the Aurignacian people of Europe from the Upper Paleolithic period first domesticated dogs. Fine jewelry and tools, often decorated with depictions of big game animals, characterize this culture.

The study explained that the scientists analysed 117 skulls of recent and fossilised large members of the *Canidae* family, which includes dogs, wolves and foxes.

> DNA studies determined all of the canids carried "a substantial amount of genetic diversity," suggesting that past wolf populations were much larger than they are today.
>
> Isotopic analysis of the animals' bones found that the earliest dogs consumed horse, musk ox and reindeer, but not fish or seafood. Since the Aurignacians are believed to have hunted big game and fished at different times of the year, the researchers think the dogs might have enjoyed meaty handouts during certain seasons.

> Germonpré[9] believes dog domestication might have begun when the prehistoric hunters killed a female wolf and then brought home her pups. Recent studies on silver foxes suggest that when the most docile pups are kept and cared for, it takes just 10 generations of breeding for morphological changes to take effect.
>
> The earliest dogs likely earned their meals too.
>
> "I think it is possible that the dogs were used for tracking, hunting, and transport of game," she said. "Transport could have been organized using the dogs as pack animals. Furthermore, the dogs could have been kept for their fur or meat, as pets, or as an animal with ritual connotation."

The history of the dog would not be complete without referring to historical evidence of the relationship between dog and man.

In March 2013 there was a study published in the *PLOS ONE*[10] scientific journal that revealed, according to lead author Dr. Robert Losey talking with Discovery News:

> Dog burials appear to be more common in areas where diets were rich in aquatic foods because these same areas also appear to have had the densest human populations and the most cemeteries . . . .
>
> If the practice of burying dogs was solely related to their importance in procuring terrestrial game, we would expect to see them

> in the Early Holocene (around 9,000 years ago), when human subsistence practices were focused on these animals. . . . Further, we would expect to see them in later periods in areas where fish were never really major components of the diet and deer were the primary focus, but they are rare or absent in these regions.

Losey,[11] went on to report that researchers found that most of the dog burials occurred during the Early Neolithic period, approximately some 7,000 to 8,000 years ago, and that, "dogs were only buried when human hunter-gatherers were also being buried."

> I think the hunter-gatherers here saw some of their dogs as being nearly the same as themselves, even at a spiritual level. . . . At this time, dogs were the only animals living closely with humans, and they were likely known at an individual level, far more so than any other animal people encountered. People came to know them as unique, special individuals.

While not expressly mentioned by Losey, there is the strong implication that the relationship between humans and dogs all those thousands of years ago was as close and intimate as many of us experience in modern times. An implication supported by authors Joshua Glenn and Elizabeth Foy Larsen in their book[12] *Unbored: The Essential Field Guide to Serious Fun* who write, "c. 10,000 B.C. In what is now Israel, a puppy is buried cradled in the hand of a human. It's the earliest clear evidence we have that humans and dogs, which two thousand years earlier had been domesticated Asian wolves, share a special bond."

If dogs had already been part of our lives for some considerable period by the time we humans had turned away from hunting and gathering, and started settling down as farmers, then we are led, inexorably, to the question of how did it all begin? What were the circumstances of early man befriending a wolf or two, and how did that very early relationship evolve into the magnificent domesticated animal that is the dog of today?

Clearly, there is no way of knowing what happened. All we can do is to dream about how that first coming together between human and wolf might have happened.

Chapter 2

THE DREAM

Neith stirred. Something had roused her from her sleep. Even with her eyes still closed she knew there had been an unfamiliar sound. A sound from the deep night outside, yet a sound that she sensed had come from close to the entrance to their shelter.

She could hear Gal's steady breathing beside her. He was still fast asleep. She opened her eyes and lay very still, seeking comfort from the familiar feel of Gal's arm across the skins that covered both of them. What was it that she had heard?

Then Neith heard the sound again. Much more clearly this time. A sound that spoke to her; spoke to her of a creature in pain. It was the sound of an animal quietly whimpering.

She softly shook Gal's arm. He woke instantly. It was instinctive. His and Neith's survival, as with all the members of their group, depended on always being alert to danger. Always keeping ahead of the many wild beasts that wouldn't, and often didn't, hesitate to feast on the unwary, or on the sick, or on their young.

Neith placed her fingers over Gal's lips, signalling him in the dark to stay quiet and listen. There it was again, that whimpering sound. The sound of a very scared, small animal. Or possibly the sound of more than one animal.

The darkness of the night, their shelter of thick grasses and reeds surrounded by the open savannah, made it impossible for them to leave just now and seek out the injured creature. There was nothing for it but to wait for the sun to rise, for the sun to light up the sky and shine down onto the land.

They sat back-to-back, their bedding skins around them, each listening. Each trying to identify the animal, or animals, from the sounds. Then there was the stirring of a faint nighttime breeze, the gentle air wafting across their entrance. The breeze carried a familiar odour. Gal picked up the unmistakeable scent of wolf. Not an uncommon odour because the wolves were constantly shadowing them, drawn by the smells of their food, hoping to find a scrap of meat, a bone, or a piece of skin. But Gal could not understand. These were not the sounds of a family of wolves; this was not the smell of a group of wolves.

Slowly the blackness of the night sky gave way to a hint of pale seen at the edge of the land from whence the light of day always came. The paleness spread and became half-light. Neith left her shelter and visited the six other shelters that made up their group. One by one, she quietly entered each shelter and, almost silently, touched each sleeping woman and man on the shoulder or arm and, as they stirred, motioned to them to remain perfectly quiet. Each of them in turn smelt wolf, heard the whimpering, and knew they must wait for more light.

Then it was time. Time to search out the cause of the whimpering, to understand what was out there. Four of them left their shelters. Gal with Neith, and Sanga with Turgunn, both experienced tribe elders, especially when it came to dealing with the animals who preyed on their peoples. All four of them fanned out and, as quiet as that morning breeze, slowly followed the scent upwind.

It was not far to go. As they closed in on the sounds of pain, it became clear that not only were there two wolves, but most likely there was a wolf of each gender. They all knew from past experiences how the sounds of a male wolf sounded so differently from that of the female.

Then they saw them. Just a few strides away, two young wolves perhaps of age only a little more than the passing of a single moon. The two pups had been attacked by an unknown predator and the rest of their family must have abandoned them.

The tearing of their small bodies was clear; blood all over their fur. The two frightened animals became quiet as the four of them approached. There was nothing that could be done. The pups must be left because it would only be a matter of time before more predators arrived to take advantage of an easy kill.

Yet there was a spirit speaking to Neith that motioned her forward. Moments later, she was crouching next to the shivering creatures. These two young wolves were so utterly exhausted. Too tired to move, unable to flee to safety. Now Neith was speaking quietly to them, soft loving tones in her voice. She sensed that deep in the minds of these tiny animals, there was a spirit whispering back to them. That they knew that Neith was not coming to harm them. That this animal who walked on two legs, who made sounds like no other animals in the land, was going to help them.

Neith's arm slowly reached out and the fingers of her hand drifted across one of the tiny heads, the gentlest touch of a human finger on the fur of this one young wolf. The whimpering stopped. The pups became very still. Neith knew what she had to do.

Neith loved the young wolves in a way that she would not have understood before that day when the animals were first found. She was unsure as to why receiving these wild animals into their midst had flowed so calmly; like the clear stream from where their group took its water. Yet it had.

The wounds had been cleaned and treated in the same way that Gal would treat a wound he might suffer when out hunting or gathering food. There had been no resistance to Gal's caring attentions. The puppies soon had readily accepted the same meats and drunk the same waters as the rest of them.

Yet Neith had this sense that these young wolves were unsure of their new lives.

That is until a dark, stormy night about three moons later on.

Neith and Gal were asleep, as were all the others in all the shelters. Suddenly, yet not alarmingly so, Gal was awakened by the soft touch of a nose on the cheek of his face. He felt the cold wetness of a wolf's nose. It was the nose of their young, male wolf. Gal did not understand how but just knew that the wolf was warning of a danger to them all.

He threw off his skins and stood. The wolf led Gal to the entrance to the shelter. The second pup then came up beside Gal and immediately both wolves opened their jaws, nostrils flaring, and let out loud howls. There was the strongest sense that the wolves were warning off an unknown creature close by in the dark.

By now, the howling of the two wolves had all the people awake in the other shelters. All the men were at the shelter entrances. Some were holding clubs and spears but all were shouting at the creature that had approached the camp. For they all trusted the instinct of their young wolves.

Then the wolves ceased their howling and the men knew the danger had passed and the morning would tell all.

The morning did tell all.

For in the soft earth around the camp they found paw prints. Prints that told of two large mountain cats that could prey on their people, especially at night when they were all sleeping.

Later on the elders sat out in the warm sunshine speaking of what had happened that last night. It was decided. Their wolves would be welcomed as friends, as new members of their family. The wolves were free to stay. The wolves would be given names chosen by Neith.

So it came to be. Neith named the female wolf Gula and the male wolf Kalbu.

Then one night, when less than a single moon had passed since that warning of danger, Neith turned in her sleep, her warm back now against Gal's back, something so common for them at night. But this night, Neith had this knowing, even in her sleep, that the wolf spirits had spread magic over their people. For alongside her lay Gula and Kalbu. Curled together sharing their sleep with Neith and Gal.

As is the way of things, as has always been the way of things, Kalbu and Gula soon settled into the patterns of the lives of Neith and Gal and the rest of the group.

The sun became high in the sky and the fruits of the warm days were bountiful.

Gula and Kalbu were very smart wolves. They knew so well where food could be found. They seemed so happy hunting with the men and, in turn, the men knew they were all hunting so much better than ever before.

These were good times as the long, warm days moved over and the shorter, cooler times came to pass again. Then the blossoms and the grasses were sending signs that the longer, warmer days were returning.

Neith was devoted to Gula and Kalbu and not a single day or night passed without many hugs and licks and kisses and strokes. So it was clear from the start that Neith knew that Gula was carrying babies. Gula became more round and soon the mother wolf was spending her time curled up on the sleeping skins, happily taking food from Neith's fingers.

Then early one day, when the sun was just above the edge of the land, Gula gave birth to four beautiful baby wolves. These new children were born into the world of the two-legged animals.

And forever more this would be the way of the human and the wolf.

Chapter 3

A GERMAN SHEPHERD NAMED BOY

My father, Frederick William Handover, was an architect, or more accurately put, my father was both a chartered architect and a chartered surveyor, which I am led to believe was quite an achievement.

At the time of my birth, in November 1944, my father was chief architect at Barclay Perkins. That is unlikely to mean anything to anyone unless I explain about the Anchor Brewery. This was an English brewery located in Park Street, Southwark, London, just a stone's throw from the banks of the River Thames. Not only had the brewery been established in 1616 but from 1781 it was owned and operated by Barclay Perkins & Co., that at one time was the largest brewery in the world.

Early in 1956, a new public house, or pub as they are colloquially known in England, that had been designed by my father, was close to being ready to receive its first landlord. It had been named The Jack & Jill by the brewery and was located in Coulsdon, Surrey, just a few miles from Croydon to the south of London. Barclay Perkins had selected a couple, Maurice and Marie Davis, who had previously been running The Orchard Hotel in Ruislip, Middlesex, to be the new landlords.

Like many in the hospitality trade, Maurice and Marie had a German Shepherd dog. His name was Boy. Again, in common with so many others in the business, they struggled to take vacations together, even short ones. So the transition from The Orchard Hotel to The Jack & Jill gave both of them a golden opportunity to take a good long holiday. However, the last thing they wanted to do was to put Boy into a kennel for a number of weeks.

My father, who clearly must have been a lover of dogs, offered to have Boy come and stay at home with us while Maurice and Marie were going to be away. Home was in the quiet street of Toley Avenue, near Preston Road, a few miles from Wembley in northwest London.

So it came about that Boy arrived at the start of the long 1956 summer school break and quickly settled in to his new surroundings. My mother, who was more of a cat person, and my sister, Elizabeth (then only eight-years-old), were somewhat overawed by this very large, albeit friendly, dog newly arrived in the household. Thus it fell to me to look after Boy and I needed no encouragement to so do. I was very quick to play with him, stroke him, and to cuddle up to him. Indeed, right from the start of Boy being in the house he slept in my bedroom. In quite a short time, I was taking Boy for walks on Barn Hill, a public area of open grassland and trees less than a quarter-mile from our house.

Sixty years later, give or take, as I sit here in front of this keyboard, I can still see in incredible detail that short walk up to Barn Hill. The hundred yards or so down to the bottom of Toley Avenue, crossing over Ravenscroft Avenue and then walking a couple of hundred yards up to the end of Ledway

Drive. From this point there was a pathway to Barn Hill via a footbridge that took walkers over the Bakerloo line to Stanmore, one of the London Underground lines, and then a few yards later the wide, open grassland that was on the lower western flanks of the hill.

Boy was always perfectly docile and restrained on his leash, especially bearing in mind that weight for weight he was almost the same weight as this eleven-year-old in charge of him. Once off the leash, upon reaching the grassland, Boy never got into trouble, nor did he ever run off and, without hesitation, came back to me when it was time to walk home. In fact, I recall all these years later that Boy would even stare with interest, without a hint of anxiety, through the open metal fencing to either side of the footbridge when our crossing coincided with a Tube train passing underneath.

Well, that never running off was a slight exaggeration. For towards the end of that July, in 1956, I was out walking on Barn Hill with Boy one hot afternoon. The weather was unsettled but not untypical for late July. From out of nowhere there was a rumble of thunder and Boy took off and raced away from me. I last saw him charging up the facing flight of the footbridge steps, disappearing down the far steps and that was that.

As quickly as I could, tears streaming down my face, I started running for home, convinced that I had lost Boy, that he had run off in his panic and become lost somewhere unfamiliar to him. But joy of joys, when I reached home there was Boy sitting on the front doorstep with an expression on his face that suggested I had taken too long a time getting myself home.

Those precious weeks in 1956 left me with life-long, unforgettable, enchanting memories of one dog, of Boy, and an enduring love for German Shepherds. Memories that were soon to become very poignant.

For some four months after Maurice and Marie came to collect Boy and take him to his new home at The Jack and Jill, my father was dead. He died from the effects of lung cancer in the early hours of December 20, 1956. I had turned twelve years old just six weeks previously.

Chapter 4

A GERMAN SHEPHERD NAMED PHARAOH

I stood very still as Sandra approached me with the golden-brown puppy in her arms. The puppy was exquisite. A fully formed yet miniature version of the adult German Shepherd dogs that were all about me here at Jutone, Sandra's breeding kennels situated at Hennock on the eastern flanks of Dartmoor in southwest England.

Sandra passing puppy Pharaoh to me, Sept. 2003.

It is September 2003 and an unusually warm day. I had unbuttoned the cuffs of my blue and white, checked, cotton shirt and folded the sleeves back above both elbows. Sandra offered me the young puppy and I took him tenderly into my arms and cradled the gorgeous creature against my chest. The pup's warm body seemed to glow through his gleaming fur and for me this moment of first contact was pure, absolute magic. As my bare forearms touched the soft flanks of this quiet little creature something registered in my consciousness in ways that couldn't be articulated but, nonetheless, were as real as a rainbow might be across the Devon hills.

There was no question that this first contact was a strong experience for both man and dog. For even at the tender age of twelve weeks, this tiny dog appeared to sense that this human holding him so longingly was deep in thought, far away in some remote place, emotionally trying to bridge a divide of many years, all the way back to those few months with Boy in 1956 and the connection with my late Father.

Behind me there was a wooden-slatted bench. I went over with the puppy still cradled against me and sat very carefully down. Now I could rest this beautiful animal on my lap. He was adorable. Large, oversized ears flopped across the top of a golden-brown head. That golden-brown fur with countless black hairs intermingled within the predominantly tan tone of his coat flowed across the shoulders, then morphed into the cream colour of his soft, gangling front legs. The puppy Shepherd almost purred with contentment, his deep brown eyes gazing so intently into my own eyes. I noticed that his eyes were softening, maybe just a hint of eyelids starting to close.

I had never before felt so close to an animal as I did just now. In a lifetime of more than fifty-nine years, including cats at home when I was a young boy and a pet cat when my own son and daughter were youngsters, I had never, ever sensed the stirrings of such a loving bond as I was now experiencing. As the young puppy seemed to be sensing in return. This was going to be my dog, without a doubt.

"So, Sandra, tell me again what I need to know about raising a German Shepherd."

"Well, we've discussed his feeding needs, so that's a big step. At first just offer him lots of care and love so he quickly registers that your home is his home. Shepherds are very bright, instinctive animals. Just look at the way that he is watching your face right now."

Sandra and I paused in our conversation to register the intentness, for that's what it seemed, in the puppy's face as he looked up at me.

"Once you have him home, Paul, start into a routine in terms of potty training. Let him out into your garden in his puppy harness so that he can sniff around. As soon as he takes a pee or a dump reward him with kind words, a rub between the ears, even a small biscuit. He will very quickly learn to potty outside."

Adding, "You know I'm only a phone call away if you have any queries."

Thus it was on this warm, sunny day in September that I drove the twenty miles from Jutone Kennels in Hennock back via Bovey Tracey to my Harberton home just a few miles west of Totnes. The little pup, as quiet as a mouse, was curled up on a blanket inside the puppy carrier placed on the passenger front seat. The passenger seatbelt around the front of the carrier, just in case.

Once in the village of Harberton, I made my way to the end of the short cul-de-sac that led to my house. I turned right into the driveway and came to a halt. Getting out of the car I walked back and closed the wooden, five-bar gate before returning to the car and driving the short distance up the gravel drive and parking the car. I opened the passenger door, lifted out the puppy carrier, and set it gently down on the warm grass. A couple of weeks before today the garden fencing had been double-checked.

A soft, wet nose led the rest of his small body out of the carrier, a nose cautiously sniffing and smelling the blades of grass about him. He padded across to a small tree, squatted and had his first pee in his new home.

Neighbour Michael, who had seen us return home, came up to the closed driveway gate and called out: "So you got him, then!"

"Hi Mike, yes we have just got back."

My little puppy, his eyes glistening with curiosity, came over to the foot of the gate and sniffed Michael's shoes that were just within reach.

"Oh, he's a cute little fella'. Did you have difficulty choosing him?"

"No, Michael, not at all. Sandra, the breeder, only had three puppies that were available just now and this little lad seemed to bond with me, and me with him, in a way that just wasn't echoed with the other two pups."

"Plus, I always wanted a male Shepherd and the other pups were both females."

Michael put his fingers through the gap between the lower two bars of the gate. The puppy lifted his head and sniffed and licked Michael's fingers.

"What are you going to call him?"

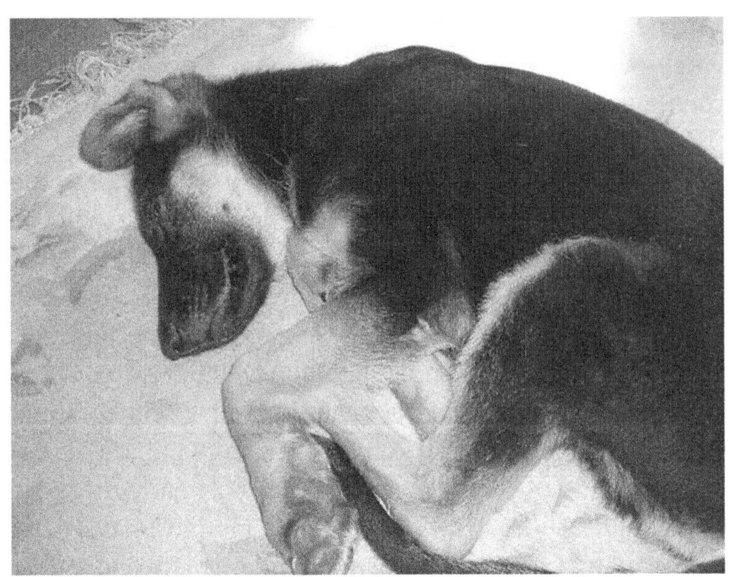

Puppy Pharaoh just 3 months old.

I responded without hesitation: "Pharaoh".[13]

Despite this being the first time I had taken on a pet dog, and a new puppy at that, Pharaoh settled down ever so easily at home. My wife, and her young daughter who was part of the family, also had very little experience with dogs but, taken in the round, Pharaoh became our new family member with a minimum of upheaval. Even puppy Pharaoh's chewing phase passed by without too many disastrous events.

The days slipped into weeks with young Pharaoh becoming such a wonderful part of daily life that soon it was very difficult to imagine my world ever being without him. So much of my day seemed to revolve around looking after this young dog. I was fortunate that just a little over five miles away from home my nephew, Lawrence, had thirty acres of land near Staverton, just the other side of Totnes, included in which were fifteen acres of the ancient Staverton Woods running alongside the River Dart. Even better, the entire property was surrounded by a stock-proof fence.

So almost from day one, I would put Pharaoh into the back of my old Volvo Estate and drive across to those secluded acres of woodland for an hour or two of exploring all the smells that Mother Nature could offer a young dog. Indeed, by the end of October it was a routine that each of us looked forward to immensely. Pharaoh would busy himself in ways that only a dog can do, totally lost in his world of trees. I would settle myself down on an old stump and just let this wonderful dog have the time of his life. Already the close bond with my young Pharaoh and, in return, his clear devotion to me, was giving me something I had never previously experienced: unconditional love.

With the early signs of the approaching winter now unmistakable, I was conscious that I was leaving it a little late to sign up for dog training classes. In many ways understandable as Pharaoh was learning so quickly and naturally from me, almost as though within a few weeks he could interpret my words and intentions. There was no doubt that he was listening intently to what was being spoken by me and to my body gestures. For instance, he quickly learned the meaning of sit, stand and lie down and then within just a couple of weeks Pharaoh knew that when I said casually to my wife, "Guess, I better take Pharaoh for a walk," he would get so over-excited that I amended saying the word "walk" to spelling it out w-a-l-k. But within days of that change, Pharaoh had learnt that spelling out the word didn't change the intention, and his excitement returned. Nevertheless, the time for training was now if I was to take Pharaoh anyplace where there would be other people and dogs, for Pharaoh was rapidly losing his puppyhood and growing into a significant male German Shepherd.

I rang Sandra Tucker at Jutone and she recommended the South Brent Dog Classes, just a few miles away from Harberton. Thus on that first Saturday afternoon in November, grey clouds spilling down from the flanks of Dartmoor, a hint of drizzle in the air, I drove west out of Totnes along the Ashburton Road towards South Brent. Pharaoh instinctively knew something different was happening this day despite him being put in the back of the car every time he was taken for a walk.

The road meandered out of Totnes through green country hills where the sheep population far outnumbered the human one. Totnes itself was surrounded by hills and dales as well as acres of green grassland, the latter frequently closely cropped by sheep. Every fold in those hills seemed to hold either an ancient wood or an ancient village that still felt strongly connected with the long-ago settlements that preceded these modern times. Village names such as Berry Pomeroy, Stoke Gabriel, Dartington, Aspringtion, Diptford, Rattery, and Littlehempston offered many echoes of times long gone.

I mused about the history of Totnes, about the ancient legend of Brutus of Troy, the mythical founder of Britain, first coming ashore here. Presumably, I guessed, because the town is at the head of the estuary of the River Dart and the Dart is one of the first safe anchorages along the northern coastline of the English Channel as one proceeds up from the southwest tip of Cornwall.

In fact, set into the pavement of Fore Street in Totnes is the "Brutus Stone". It's a small granite boulder onto which, according to that legend, Brutus first stepped from his vessel, proclaiming, "Here I stand and here I rest. And this town shall be called Totnes." I had frequently pondered that the likelihood of the legend being true was pretty low, but it was a great tourist magnet!

Just six miles later, Pharaoh still sitting erect intently watching the passing cars, I drove across the flyover that spanned the main Exeter to Plymouth Road, the A38. Seemingly always busy, whatever the time of day, or day of the week, the speeding cars were throwing up a road spray as the drizzle had now deteriorated into steady light rain.

I turned onto the B3372, that meandering country lane that ran into South Brent. I had been told to watch out for a five-acre field on the right-hand side just before entering South Brent, where the classes were held come all weathers.

The open field gate and half-a-dozen parked cars made the location obvious. I drove carefully in, parked on a gravel parking area, leaving some distance from the smart, white Ford van to my right-hand side, and turned off the engine.

The ignition key had hardly landed into my coat pocket when Pharaoh erupted into a frenzy of barking. Thirty yards away, a cheerful Cocker Spaniel was being walked across to the gathering group of dogs and their respective owners, and this clearly had triggered the barking.

Pharaoh's nose was pressed up against the tailgate glass, his whole body tense, ears erect, and tail straight out. This was a dog in an attack posture. The sound of barking was overwhelming in the confined space of the car.

"Pharaoh! Shut it. Quiet!" I shouted.

Pharaoh stopped barking but was still quivering all over, giving every indication of wanting to jump out of the car and beat up the Cocker Spaniel.

This was not what I had anticipated, far from it.

I swung my legs out of the car, stood up and closed the driver's door. I needed to find the person in charge of the class and get acquainted with the routine. The rain was typical for Dartmoor. Fine rain that seemed to have a way of working its way through the most tightly buttoned coats. I pulled my coat collar up around the back of my neck and walked across to where a group of people were standing together, perhaps half of them with dogs held close to them on leashes.

As I approached the group, a woman, perhaps early middle-age, her dark-brown hair spilling out from under a leaf-green felt hat, caught my eye. She walked over to me, her blue jeans tucked into black Wellington boots.

"Hallo, you look like a first-timer?"

"Yes, that's correct. My name is Paul and I'm from Harberton, together with my German Shepherd: Pharaoh."

"Well, welcome to the class, Paul. My name is Deborah and I'm the instructor around here. Do call me Debbie, most do."

I quickly guessed that she was an experienced coach.

"Was that Pharaoh that I heard barking a few moments ago?" she asked me.

"Yes, that's the first time he has behaved like that."

Debbie was looking across at the Volvo. "Strong, male German Shepherd, I don't doubt. Not uncommon at all," she replied, continuing, "Leave him in the car until I have the first class underway."

She went on to explain, "We have the regulars walk around that grass area over there with all their dogs on leashes. This gets them settled down. Then we reinforce the usual commands, as you will see."

I looked to where Deborah was indicating. The nearest corner of the grassland, that must have been three or four acres in size, had an area that showed clear previous signs of dogs and owners walking round in a wide circle.

"After that, in about twenty minutes," she continued, "then we will bring Pharaoh in with, perhaps, just two or three other dogs, and see how he behaves."

Then adding, "Once the first class is running, I suggest you give Pharaoh a bit of a walk over in the far part of the field, away from the others, just to get him adjusted to the environment."

"Oh, and I presume Pharaoh is settled on the leash?" she added as an afterthought.

I replied, "Yes, he's fine on his leash. In fact, he walks well with me despite no formal heeling lessons."

Debbie responded, "Shepherds are incredibly intelligent dogs and I can tell just from the way you speak about him that the two of you are very close. Catch you later, must dash now."

I went back to the car and reached in to the rear, pseudo-leather, bench-seat and picked up Pharaoh's leash. It was a handsome affair, even if was just a dog leash. Sandra, from the breeding kennels, had recommended the type, a leash that had two length settings. More important, the supple, heavy-duty leather leash had a hand loop just six inches up from the snap catch. This allowed me to hold the leash in my left hand with Pharaoh having no freedom to be anything other than close to my left leg, the recommended arrangement for walking a dog to heel. Right from the first moment that Pharaoh had been taken across to Lawrence's woods, I had taught him to "heel" on the leash as we walked the grassy track down to the woods. It wasn't long before Pharaoh would obediently remain close to my side without any pulling, even with the leash at full length. But how would it be today?

I leaned over the back of the bench-seat and clipped the leash onto Pharaoh's collar before slipping back out from the car and closing the side door. I walked around to the tailgate and inched it open, sufficient to slip my arm inside and grab the leash. With my other arm, I raised the tailgate to its full extent. Pharaoh sat on his haunches just staring at everything about us.

"Pharaoh, down you get, there's a good boy."

He effortlessly dropped down on to the grass and looked up at me. His overall demeanour signalled that these were first-time experiences in this young Shepherd dog's life.

I gave him a couple of quick commands, "Pharaoh, stand. Pharaoh, heel."

With that I stepped, left foot forward, and Pharaoh immediately dropped into formation, as it were, alongside my left knee.

It was a walk of perhaps a couple of hundred paces to get to the far corner of the field. The ground had risen in this direction and now when I had Pharaoh sit and I looked back across the field and beyond, the view was rolling South Hams countryside so typical of this part of South Devon. Even with the light rain, this view was comfortingly homely.

Despite a lifetime of living in so many different places, both within the United Kingdom and overseas, this part of Devon felt strongly connected to the person inside of me, that this was my home, this was where my roots were. That Acton, my place of birth in North London, just happened to be a technicality in my life's journey.

Before we knew it, the initial group was leaving the walking area and it was time to experience Pharaoh's first obedience class.

We waited just to one side. Debbie came across.

"Paul, do you know what Pharaoh is like with other dogs?"

"No experience whatsoever," I replied, continuing, "We live over at Harberton but I have access to private woods at Staverton. Pharaoh is walked there most days. So I have never walked him in a public place and had no intention of doing that until he's been properly trained and assessed, by you I guess."

"OK, let's take it cautiously. Will you walk Pharaoh into the centre of the exercise area, have him sit, and hold him close to you on his leash." She was quiet in thought for a few moments, unaware, it seemed, of the rain water that looked as though it was soaking into the crown of her hat.

She added, "We have many dogs here today, although no Shepherds. I will ask a few of the owners to walk their dogs, dogs that I know well and can trust, one at a time in a circle around you and Pharaoh, coming in closer each time if it all runs to plan."

I walked Pharaoh to that centre spot. "Pharaoh, sit." He did so without hesitation.

A black, female Labrador and her owner, a gent wearing a navy-blue, full-length raincoat over brown hiking boots, his right hand carrying a wooden walking stick, detached themselves from the group of dogs and owners and commenced to walk obliquely around me and Pharaoh.

I reinforced my instruction to Pharaoh with what I hoped was a strong tone in my voice: "Pharaoh, stay. Sit. There's a good boy."

Debbie was thirty feet away watching the proceedings carefully.

"Tom," Debbie called out to the circling gent, "come in just a few more feet and continue circling around Pharaoh."

"Thanks, Tom, that's fine. Geoff, would you be next?"

Tom and his Labrador returned to the owner's group and Geoff, a younger man, perhaps in his late twenties, came over with a smaller, creamy-coloured male dog. To my eyes the dog looked like a Pit Bull or a Pit Bull mix.

This dog, however, was a far less settled creature than the Labrador, and Terry, for the name of the dog immediately became clear, was prompted several times to stay close and heel.

Terry and Geoff approached the circling zone. Immediately, I became aware of Pharaoh starting to bristle, the hairs on the nape of his neck lifting in anticipation of something, something only known to Pharaoh.

"Pharaoh, sit," I voiced sternly as I noticed his rear quarters just lifting up from the wet grass.

As Geoff and Terry circled around the rear of us, Pharaoh squirming his body and head so as to keep an eye on this other dog, it happened.

As the Pit Bull arrived off to my right side, perhaps about eight feet away, Pharaoh sprang at the dog. It was not entirely unexpected but even so, even with Pharaoh being held at short rein, the jump practically dragged me off my feet. I had no idea that Pharaoh now had such power in his legs. He was just a little over four months old.

"Pharaoh, no! Come here! Come back!" I shouted angrily. Then continued shouting commands while at the same time dragging Pharaoh back to my left-hand side. He begrudgingly obeyed but continued barking fiercely, standing erect on all four legs, lips curled back exposing his fangs and teeth; everything about him signalling to the Pit Bull that Pharaoh was a deeply, unhappy dog.

Debbie signalled to Geoff to retreat from the area and, as quickly as Pharaoh had become upset, he settled down and squatted back on his haunches.

Debbie walked across to me.

"Paul, I'm terribly sorry to say this," she said quietly, "but I think you have a German Shepherd with an aggression problem. Until you get that sorted out, I just can't take the risk of Pharaoh coming to these classes. Under the circumstances, I'll waive today's training fee."

With that Debbie turned on her heels and returned to the other owners.

Chapter 5

PHARAOH, THE TEACHING DOG

I was gutted and utterly shocked to my core. The dog that meant so much to me had been rejected. That rejection felt as much my rejection as it was Pharaoh's, and it hurt!

Our drive back home to Harberton was altogether a different emotional experience from the one when we had earlier left for South Brent. I just couldn't get my mind around what had happened and questions boomed around my head. Why that one incident had branded Pharaoh as a dog with an aggression problem, why the trainer hadn't been better prepared, why there seemed no sensitivity to this being such a new experience for me, and on and on. But as much as the thoughts kept running around my mind, it didn't in any way alter the fact that I hadn't a clue as to why Pharaoh had behaved in that fashion, and where next this was all going.

Accepting that this was the first time I had ever owned a dog and that I had no experience of being a dog owner, my close bond with Pharaoh convinced me that there were no dark behavioural issues that needed dealing with. Mind you, that only left me feeling even more confused.

I turned right off the road that ran from Totnes to Harbertonford into the small lane, high-sided with tall hedgerows that dropped down into Harberton. Less than a mile later, I was turning into the short driveway in front of the house and parking in my usual place. Leaving Pharaoh in the Volvo, I walked back down the driveway and closed the wooden entrance gate.

Pharaoh jumped down as soon as the Volvo's tailgate was raised. The one, small, positive thing was that it wasn't raining. Pharaoh sniffed around, cocked his leg against the stone wall that fronted a raised flower bed then skipped up the four stone steps, across the gravel in front of the house and waited for me to open the front door.

I took off my raincoat and hung it on the hooks at the rear of the hallway then walked up the wooden stairs that led from the level of the front door to the living-room on the first floor. In short order, Pharaoh was up the stairs and immediately settled himself in front of the black-iron wood-stove in the corner of the living room, hogging the warm glow that still radiated out, me having made up the stove before we went out earlier on. I made myself a tea in the kitchen, walked over and sat down on the settee. I knew that deep inside of me I felt emotionally ripped apart.

I must have been transmitting my angst at the way the morning had turned out, for Pharaoh raised himself from the fireside carpet, gently came over and softly, almost tenderly, laid his head on my right thigh. There was no other way to describe this than pure, unconditional affection. A simple, yet beautiful, symbol of love by a dog for a human. It turned out to be much more than a symbol because it brought me out from my self-despair.

CHAPTER 5

The next day, a Sunday, I awoke a little before eight in the morning, and despite the weather still being poor with low grey clouds scudding overhead and the ever-present threat of rain, I washed and dressed, made myself a quick breakfast, grabbed Pharaoh's leash, the keys to the Volvo and headed down to the front door. Pharaoh, of course, had already guessed it was walking time, despite it being earlier than usual. He bounded out of the front door and down the few steps to the driveway and waited expectantly for the car's tailgate to be opened.

Twenty minutes later, I was walking him down the grassy track of Lawrence's large twelve-acre field that stretched out to my left, a dark hedgerow to my right, and Staverton Woods a couple-of-hundred yards ahead of us.

This tiny paradise deep in the heart of South Devon meant so much to me. Cut off from people, phones, the Internet and all the consumerism of modern life, this was the place where I could restore some form of mental balance. I often wondered about what these lands could tell if only the ancient pastures and woodlands could voice their histories. Staverton Woods was known to be very old and when Lawrence was bidding for the property he only managed to win it by a nose from the Woodlands Trust that was going to preserve the woods for evermore.

But Lawrence and John, his dad, had been equally fine stewards. The woods were still unchanged from long, long ago. All that Lawrence had done was to convert three acres of the top grassland into a large bed for the planting and harvesting of eucalyptus trees. There was a ready market for the branches in the floristry trade.

In the springtime, the woods were glorious. The mix of larch, ash, and old species of oak trees that can only come from years and years of being left untouched was also blended with bluebells, the dainty blue flowers practically covering the ground beneath the trees, underlining why the locals called them Bluebell Woods rather than Staverton Woods as depicted on the UK Government Ordnance Survey maps.

Pharaoh, released from his leash, bounded off to check out once more whatever it was that he checked out each time we came here. Meanwhile, I slowly worked my way into the depths of the woods.

The long, steamy, sound of a locomotive whistle suddenly echoed through the air. That was not uncommon, for at this point the line of the Dart Valley Railway ran alongside the trees, sandwiched between the edge of Lawrence's woods and the bank of the River Dart.

The line, running between Paignton and Dartmouth, had been a victim of government cuts, the so-called Beeching cuts back in the late sixties, but had been rescued by a then newly formed Dart Valley Railway company and operated successfully ever since. The chuff-chuff-chuffing sound of the black steam engine, smoke and steam rising into the damp valley air, heralded a train consisting of three cream-and-brown passenger coaches. It so perfectly matched a sense of earlier, simpler times for the railway had been completed, if I recalled correctly, way back in the mid-eighteenth century.

The rear of the last coach, sporting a pair of the red-lensed oil lamps, disappeared from sight around the bend of the riverbank. I returned to my thoughts.

I had woken that morning conscious that my instinct about Pharaoh's nature had hardened into being certain that

Pharaoh looking out from a Dart Valley Railway carriage window.

Debbie's analysis of Pharaoh was utterly wrong. I would stake my life on the fact that Pharaoh was not an aggressive dog.

Nevertheless, as I stood here under the trees, I had to admit to myself that Pharaoh had acted in a way towards that Pit Bull that, at the very least, appeared to be antisocial.

What to do? What on earth to do?

Then it came to me. Pharaoh needed to be observed with other dogs in a less stressful situation than that of yesterday's obedience class. How about me walking him on Dartmoor? It was a Sunday morning, not unreasonable weather for the time of the year, and there would be plenty of walkers with their dogs out on the moor.

I called Pharaoh back to me, snapped the leash on to his collar, and walked him back to the car. By the time we had reached the car, a further idea had come to me. An idea prompted by that view of the River Dart a short while ago. That was that I had always meant to find the source of the River Dart. I knew it was somewhere up on Dartmoor but in all my years of living in South Devon I had never taken the time to find the exact spot. I would first go to Dartmeet, named because that was where two branches of the young river came together, a favourite place for walkers with lovely pathways along the riverbanks. In fact, this was turning out to be a brilliant idea as the back road from Staverton, across the A38 Exeter to Plymouth road, and on up to the moor, more or less followed the course of the River Dart.

I started the engine and reversed carefully out of the field entranceway back into Sandy Lane. I loved driving along these narrow Devon lanes, often no wider than a tractor and trailer. What fascinated me was that when two vehicles came face-to-face, each driver seemed to know instinctively who had the closest lay-by or field entrance behind them. There was never any "argy-bargy" about the issue. Except, that is, during the summer months when a visitor to this part of the world tried out one of the lanes, or got lost. Then it was a case of stepping out of the car and saying to the other driver that you think the passing place is closer to them than it is to you. As often as not, it was just simpler to reverse back rather than suffer the ire of a tourist who wasn't so hot at reversing along a narrow Devon lane. Early on in my Devon days I had learnt to reverse using my wing mirrors, a practical necessity around here.

I smiled in recollection of the day when I came bumper-to-bumper with a driver who simply couldn't reverse her car. Almost immediately, another couple of vehicles had pulled up behind me so there was no choice other than the woman's car had to be reversed. She was adamant that she couldn't do it. But agreed to me sliding into the driver's seat and reversing the car for her. Luckily only reversing about three-hundred yards back down the lane to a passing place. The other drivers had been very patient, indeed seeing the funny side of the situation.

Sandy Lane became Cabbage Hill leading to the bridge over the A38 that below was as busy with traffic as usual. Practically every square inch of the land either side of these roads was cultivated or cropped grassland. Yes, it was very rural. Yes, it was a very ancient part of southwest England. But the agricultural intensity of the land, a very modern phenomenon, was unmistakable.

Once over the A38, the lane ran around the left-hand flanks of the village of Ashburton, just off to our right, and then at the top of Bowden Hill, the narrow road headed more or less directly, or as directly as any Devon country road ever did, towards the southeastern flanks of Dartmoor. A few miles later, at the top of Newbridge Hill, just a quarter-of-a-mile after passing through the tiny hamlet of Poundsgate, the road forked. My route took me to the left and I had just started the turn when, quite suddenly, I noticed out of the corner of my eye a sign hanging from a tree at the start of the right-hand fork. It read: "GSD Club of Devon Meeting – This Way."

I braked to a halt and reversed carefully back the few yards to the junction. I had never heard of the German Shepherd Dog Club of Devon. This had to be investigated.

I took the right-hand fork and within moments the lane was running through heavily wooded land. We must be within the edge of Dartmoor, I speculated, because it was well known that the lower flanks of the moor were heavily forested, all protected woodlands, thank goodness.

Five minutes later, there was a further sign pointing the way to a private lane. I slowly drove up the lane and almost immediately saw a professionally painted name board: "Angela Stockdale – Dog Aggression Specialist."

I just didn't know what to think, what to feel, and had no idea what on earth was going on. I was not a believer in the traditional religious sense but also didn't label myself as an out-and-out pagan. Tended to use the term "spiritually minded" when relevant to so describe myself. Thus, was it just serendipity that had brought me here, or what!

I drove slowly into a yard surrounded by many pens and buildings, stopped the car, and stepped out. I was aware of the sounds of barking coming from a number of directions. All sounding like German Shepherd barks.

The click-clack of a metal pen gate being closed caught my attention. I looked to see a woman checking that the gate latch was closed and then turning my way.

"Hallo, can I help you?" she called. "If you are here for the club meeting then you are about three hours too early."

She walked towards me. Despite the grubby blue overalls that she wore, bottoms poked into a pair of red rubber boots, she exuded an attractive warmth. Her thick, auburn hair bracketed a pleasant face with little makeup. I noticed a blue-and-black necklace close around her neck. I surmised that this was a working lady who was still in touch with her femininity.

"Hallo, I'm so sorry to arrive unexpectedly like this. I was on my way to Dartmoor to walk my dog, chose to come the back roads from Staverton, and just happened to see the sign for the GSD meeting and followed the signs out of curiosity."

I paused. "Well, perhaps I should add a little more than just curiosity. For I have my German Shepherd in the back of the car and just yesterday at the South Brent obedience class he was accused of being an aggressive dog and we were told not to return."

"Hallo to you. My name's Angela and good to meet you. Perhaps I shouldn't say this but that issue at your dog class, would that have been Debbie's class?" I nodded. "Well, just let me say that you could do a great deal better."

"Angela, wow, am I pleased to meet you. Talk about an unexpected gift." I paused before continuing, "I'm Paul, Paul Handover, and the dog in the car is Pharaoh who was born in June. We live over at Harberton just to the southwest of Totnes."

I pondered for a few moments, then said, "Look, I was on my way to the moor to see how Pharaoh behaved with other walkers and their dogs. Almost exclusively, I have been walking Pharaoh over at my nephew's woods at Staverton. So I haven't been getting him accustomed to other dogs as perhaps I should have been. Would there be any chance of you assessing him and offering me some proper guidance? I'm a first-time dog owner, you see."

"Yes, of course, Paul. That's what I do here. However, I'm not even going to suggest you let Pharaoh out now, too much going on, and just not the best circumstances for him."

She took a small spiral-bound notebook from her overall pocket, opened it, and looked through a couple of pages. "Can you and Pharaoh come here, say eleven in the morning, next Wednesday?"

"Yes, we can, without any difficulty. Is there anything that I should bring with me?"

"No, other than just bring Pharaoh's usual leash. Oh, and you might want to give him a good walk before you get here."

She seemed to be mentally checking if there was anything else, but just added, "That's fabulous, I will see you both in just three days time."

"Angela, thank you. I can't wait for you to meet Pharaoh. Oh, and good luck with your meeting this afternoon."

With that I turned and got back into the car, started the engine, swung the car in a tight circle and drove carefully out of the yard.

Glancing up at the rearview mirror, I saw that Pharaoh was looking out at Angela and realised that there hadn't been a peep from him while I had been speaking with her. I wondered if Pharaoh had been picking up the vibes of our change in fortunes. A little voice in my head answered me saying, of course Pharaoh had understood. Undoubtedly, Wednesday would reveal all.

I returned home in a very different mood to the one I had felt when we had left the house a little over three hours ago. I opened the front door, allowing Pharaoh to push past me, as he always did, and stepped into the house. I made myself a coffee and sat quietly next to the full-length glazed door that could be opened from the living room into the garden.

Outside, the grey cloud was breaking up and letting a fitful November's winter sun shine through another pair of fully glazed doors that looked southwards out over the tiny cul-de-sac where our house was situated. We had lived here for some eight years, luckily me finding Upper Barn, for that was the name of the property, at a time when I had been in rented accommodation in a farmhouse just a couple of miles away. So it was an uncomplicated move for me.

Before I knew it, Wednesday morning had arrived. Despite the fact that rationally one wouldn't expect dogs to understand the days of the week, Pharaoh's behaviour as soon as I was awake was untypical. OK, it wouldn't have surprised me to know that I was letting out some small body signs about the forthcoming meeting with Angela, but still. I quickly showered, got dressed, and ate my breakfast.

I must have become lost in thought about the subtleties of communications between a dog and a person because ten minutes later Pharaoh had to remind me, with a nudge from a warm snout, that we were going out and to, please, open that front door. A very excited Pharaoh then bounced down the steps; without doubt he sensed something very different about this day.

Again, South Devon was offering typical November weather with low clouds and the promise of rain. I had Pharaoh's regular leash plus I had grabbed his body harness that was such a gentle alternative to tugging on a dog's collar.

It was a little after nine-thirty when I parked nose-in to Lawrence's field gate. Like so many previous times, I let Pharaoh out of the car, locked the car doors and opened the gate to the upper field. Just for a change, and just as much for the experiment, once the gate was closed behind us, I commanded Pharaoh to sit.

"Pharaoh, stay." I quietly unclipped the leash from his collar. "Pharaoh, heel." I slapped my left thigh with my left hand and set off down the grassy path. As I hoped, Pharaoh trotted beautifully to heel, even up to within a few yards of the edge of the woods.

"Pharaoh, sit." I rubbed his forehead, just where the blackness of his snout filtered into the black-brown hair across his brow above his brown eyes. "There's a good boy. Go on then, off you go."

Pharaoh was away into the trees once again and I found one of my usual stumps, swept the back of my coat underneath my backside and sat down.

The next hour passed as gently as one could ever wish for and, as if on cue, Pharaoh trotted up to where I was still sitting just about when it was time to be leaving in order to be at Angela's place at 11 am.

Soon we were back in the car with me reversing out into the lane and repeating the car journey of last Sunday. I still couldn't square the circle of events since we had been marched off, figuratively speaking, from that class just last Saturday afternoon. It seemed like a lifetime ago. That old chestnut came to mind, one of many that I was apt to use: "Never underestimate the power of unintended consequences."

As we nosed once more into Angela's yard, about ten minutes before eleven, she was there expecting us. This time the muddy overalls and red plastic boots had been cast aside for a pair of freshly laundered blue jeans, that fitted snugly around her hips, worn over a pair of soft, walking shoes, all topped with a cotton blue-and-white blouse showing from under a woollen pullover. Angela's face declared more make-up than last Sunday.

"Morning, Paul," Angela called out in a bright and breezy manner as I closed the driver's door behind me.

"Good morning to you, Angela, what's the plan then?"

"It's quite simple, Paul. Just walk Pharaoh on his leash over towards that fenced-off pasture, see where I'm pointing, then stop before reaching the gate when you are about five or ten feet away."

I opened the Volvo's tailgate quietly surprised that Pharaoh was in a very contented mood. Despite the lure of so many new sights and smells, he sat quietly on his haunches as I clipped on his leash.

"Down, Pharaoh. Pharaoh, sit. Pharaoh, heel." Bless him, I thought, he's behaving immaculately.

As we came to a halt, Angela standing by the gate, I noticed that in the far left-hand corner of what was an enclosed pasture of, perhaps, a couple of acres, there were two dogs. I was totally thrown aback by her next instruction.

"Paul, I'm going to open the gate a little and stand back. Will you just slip inside the field, let Pharaoh off his leash, step a few paces away from him and then leave him to do just whatever he wants to do."

I felt rather uncertain about this and hesitantly replied, "But, er, but Angela, I can't guarantee that he won't go across and be aggressive to those dogs over in the far corner."

"Don't worry, Paul. This is not as random and unplanned as it might be coming across to you." Angela then unlatched the metal gate and opened it towards her by three or four feet. She then stood back.

Pharaoh looked through the open gate at the two dogs a good hundred yards away in the corner of the field. I stepped forward with Pharaoh and, as Angela had instructed, once inside the gate I unclipped Pharaoh's leash and stepped back from him. Pharaoh walked slowly yet confidently away from me for a farther twenty-five yards. He then stopped and just stood still. I was riveted by Pharaoh's stance and composure, as I had never seen this behaviour from him before.

Then came the call from Angela that not only made my day but was destined to remain in my memory for the rest of my living days.

"Paul, there's nothing wrong with Pharaoh."

I practically choked on getting my next words out. "Sorry? Angela, I'm not sure I heard you correctly? Did you say there's nothing wrong with him?"

Then before she had time to reply, continuing, "But I don't understand. How on earth can you tell so quickly when Pharaoh's hardly even entered the field?"

Angela came in to the field and walked up next to me. "Paul, it's very easy. My dogs are telling me that Pharaoh is

OK. Simply because both my dogs haven't taken any real notice of him. He'll be fine. Let's just go outside and lean on the fence and watch the three of them and I'll explain what's going on."

As if in a dream, a few moments later I had my arms over the top wooden horizontal rail of the fence. Angela, being a little shorter, stood next to me with her hands on the rail.

"Paul, those two dogs of mine out there in the field are Sam and Meda. They are both teaching dogs. Sam is a male that we would describe as an omega teaching dog and Meda is a female teaching dog that we describe as an alpha dog. Don't worry just now about those terms, I'll explain all later. Let's just watch Pharaoh's interaction with Sam and Meda."

I was speechless. I loved that dog of mine so much and had been so worried these past few days that to have Angela's endorsement of him in this manner was a joy beyond joy.

I watched as Pharaoh, having moved away from the spot where he had been standing, came up to the two dogs. Pharaoh's head was slightly lowered, his tail down, seemingly offering himself to Sam and Meda as a submissive youngster ready to learn.

Sam took no notice at all of Pharaoh. But Meda partially encircled Pharaoh, sniffed his bum and then came around in front and softly touched him, nose to nose. Now Pharaoh noticeably perked up as Sam came across to greet this new companion, Pharaoh's tail gently wagging a return greeting. Sam then hung back as Meda appeared to take Pharaoh on a bit of tour around that part of the field, sharing this and that smell.

"Do you know what, Paul? I'm pretty sure that Pharaoh is a teaching dog."

She continued, "The way that my Meda has taken Pharaoh under her wing, as it were, and the way that Pharaoh is being welcomed by her and how Sam is so comfortable with him all points to your Pharaoh being a minder dog."

There was a pause before she continued, "I think you have a wonderful German Shepherd, Paul. I wouldn't be at all surprised if I can't use him to teach some of the poor dogs that come this way."

As Angela spoke those last words, I burst into tears and struggled to find a paper tissue in my pocket. I apologised to Angela for losing it.

She turned to look at me and gently said, "Paul, I really understand and it's lovely to see your feelings for what is a wonderful young dog. Let's call them all in and I'll make us a nice cup of tea and open your eyes to the magical world of teaching dogs."

Angela called out to Sam and Meda and over they came with Pharaoh happily in tow. I was then able to call Pharaoh over to the car whereupon he jumped in through the open tailgate just as happy as a dog could be.

Sam and Meda had parked themselves somewhere out of sight as Angela pointed towards a static caravan that seemed to be the customer's lounge. Inside, there was a small gas burner and within minutes the kettle was singing out in the unique way that full kettles sound when they are warming up.

"Sit yourself down in the corner, Paul. Won't be long. How do you take your tea? White with sugar, or . . ."

"Just white with no sugar, please Angela. Must say that I could murder a fresh cup of tea."

"Tell you what, why don't you go and bring Pharaoh in with us here in the caravan. This story about dogs could take a while." There was a chuckle in her voice.

Moments later, Pharaoh was curled up contentedly on the opposite corner cushion. Like most dogs, he loved being in the company of humans chatting comfortably together.

Five minutes later, my fingers around a warm, white china mug, steam rising from the tea, I was all ears and ready to learn about teaching dogs. My inner voice was telling me that I was on the verge of learning something very new about dogs. It was turning into a magical morning.

Angela took a deep breath: "I guess we need to go back a very long way to get to the start of the story of dogs. As I'm sure you know, dogs are part of the species that includes wolves, coyotes, and foxes. It's a species that scientists believe evolved millions of years ago."

She took a mouthful of tea from her own mug and continued, "My understanding, and I'm no scientist, is that our forerunners who came out of Africa were smarter than the Neanderthals, that they used language and developed tools and benefitted as hunters enormously because of their relationship with early dogs."

"In fact, I have a great article from Professor George Johnson[14] who has done much research into the evolution of the domestic dog. I'll give you a copy before you leave."

I took a long drink of the tea. Gracious it went down well.

"Angela, this is fascinating and, yes, I would love to read that article. But, surely, that can't have any bearing on today's dogs?"

"Well, yes and no," she replied, going on to say, "Despite dogs these days having no instinctive awareness of the natural pack size and dynamics of their doggie ancestors they still carry the genetic imprint, for want of a better description, of the structure, the hierarchies as it were, of those ancient dogs in those ancient times."

A question came into my mind. "What about feral dogs? Surely in some countries the number of feral dogs is huge, don't they adopt the pack behaviours of the early days?"

"That's a good point, Paul, but even if feral dogs do pack together, and they do for hunting and food-seeking purposes, feral dogs are such a mixture of breeds and temperaments that there isn't a chance of a cohesive group coming together in the way that dogs did way back in earlier times when groups of dogs would have been one doggie community."

"Guess that makes sense," I reflected.

She continued, "We are pretty sure that in the early days of dogs evolving from the grey wolf, they maintained a similar social order to that of the wolf. George Johnson covers that well in his article. That is that in a pack size of about fifty animals the group was guided by just three social differentiations."

She drained the last of her tea and went on to explain, "Right from those early days there were just three dogs who had a social role, a social status as it were, in the pack. The first role was the alpha dog, always a female dog. Next in line was the beta dog, this time always a male. The final social role was the omega dog that could be of either gender. That's what was believed for years."

I reflected how in common parlance the term alpha tended to be associated with the phrase alpha male.

Angela continued, "Even in modern times, a proportion of dogs are born with those same genetic markers of status. Recently, however, it's become clear that these alpha, beta, and omega terms are a long way from being accurate. The more appropriate description is to see those roles under the general heading of teaching dogs with the additional sub-divisions of mentor, minder and nanny dogs."

She paused, "Are you following this, Paul?"

"Oh yes, absolutely, but I had no idea about any of this."

Angela responded, "Well, I'll briefly describe those differences within teaching dogs."

I noticed that Pharaoh was fast asleep and dreaming.

"Let's start with the mentor, or alpha dog in old terms. This is a dog that is normally assertive by nature, but quietly so. They are not dogs that play much, unless flirting with the opposite sex. In their position as teaching dogs they are dominant but in a way that trainers would describe as passively dominant. So they would always meet a dog with assertiveness but never with hostility. Mentor dogs relax other dogs, less with the use of body language as such, but more often because their presence just has a calming effect on most other dogs."

Angela paused, "Paul, can I make you another tea?"

"No, I'm fine. Far too engrossed with what you are saying to be interrupted by another brew-up." She smiled.

"So, let me finish off describing mentor dogs. Often the mentor dog, when working in a group of dogs, will watch from the sidelines and only become involved within the group if absolutely necessary. And, of course, that necessity is the mentor's evaluation; almost impossible for us humans to interpret. As I like to say, dogs speak dog so much better than we humans speak dog."

Precisely at that moment there was a quiet moan of contentment from Pharaoh curled up, as he still was, on the cushion. It was almost as though he was listening and approved Angela's last words. She had paused on the sound of Pharaoh's little moan and now continued.

"Mentors can be quite lazy. They have a very interesting and, to a great degree, a rather complex view of the other dogs that they come in contact with. It's a certain bet that we don't know the half of it when it comes to understanding the mentor teaching dog. For example, they will support other teaching dogs where needed showing, for instance, what to do in difficult situations if that other teaching dog is not coping. But the mental analysis and language used by the mentor dog in these circumstances is way beyond the comprehension of us humans, even those who have spent a lifetime studying dogs."

"The last aspect to mention is that there is a varied reaction from other dogs towards a mentor dog. Some dogs take great confidence in mentors and whilst not necessarily submissive towards them, they are very respectful. But others find mentors intimidating and will avoid making contact with them at all."

Angela paused.

I was blown away, to use the modern vernacular term, totally dumbfounded that there was so much more to the dog world than I could ever have imagined.

"Paul, do you want me to carry on explaining the other teaching dog roles?"

"Angela, I could listen to this all day. It's stupendously interesting."

"OK, then we have to look at the two other teaching roles that we know exist in today's dogs."

Angela continued, "The minder, or beta dog in old terms, is totally different to the mentor dog. In the sense of being different in the way they interact with other dogs. When minders meet another dog, they approach with the active intention of interacting with that other dog. The minder dog is naturally assertive, often as strongly assertive as your Pharaoh is, but ultimately not as strong as the mentor dog.

When minder dogs meet another dog in a teaching situation, they assess the new dog as it approaches and use appropriate body language in accordance to how the other dog is reacting to them. That makes them frequently more demonstrative than a mentor dog and the minder dog will actively seek interaction within a few minutes of meeting a new dog. That interaction does not necessarily mean an invitation to play, far from it. If the minders feel the other dog is not ready for that level of interaction, they will converse dog to dog in a more subtle manner."

Angela paused. "When I think about all the teaching roles, the minder dog is incredibly interesting, with so many different levels of communication going on."

"For example, if the other dog is worried but shows signs of wanting to rush at the minder, he will stand firmly with his head side on to that other dog. Eye contact is made intermittently as the minder determines whether the new dog is calming down or intending to rush at him."

"Minder dogs can also openly display assertiveness if they need to. Once the situation is under control, from the minder's perspective, the minder will generally initiate connection activities from the other dog such as by marking then walking away, allowing the other dog to investigate the minder's scent. Or, perhaps, invite the other dog into a status game, often instigating a chase."

Angela paused to sweep some hair back behind her left ear.

"Then again, if the other dog shows signs of trying to drive the minder away, the minder will turn his head towards the dog and eye contact becomes much stronger. Minders do not reposition any other part of their body. If the other dog shows signs of moving away, the minder will totally drop their assertive body language and also move away. Then the minder will reassess the other dog from a distance before possibly approaching again."

"Finally, and this is what makes the minder such a fabulous teaching dog, the minder will interrupt any unsociable or unruly behaviour. Simply by physically placing himself between the dogs in question and remaining there until the tension has reduced. Once calm has returned the minder will usually walk away and keep an eye on things from a distance. In effect, the minder is policing a group of dogs for the greater benefit of the whole group. Most other dogs recognise and respect the strength in a minder. Yet what is fascinating is that the minder, while a strong dog, does not naturally command respect in the way a mentor dog does. So you can have a situation where some dogs who have limited canine communication skills or are still adolescent can unwittingly challenge a minder."

Angela paused, noticing me ready to voice a thought.

"Bingo! Now I know what happened at that class at South Brent. I was sensing that the Pit Bull had an unruly personality and Pharaoh's reaction, I presume, was to signal that he was not welcome."

"That would have been my guess," Angela confirmed.

"So let's look at the last of the three teaching roles, that of the nanny or omega dog."

"In many ways, the nanny is the most amazing of the three types of teaching dog. Uniquely, a strong nanny can temporarily take on the role of a minder or even a mentor if needed. They are extremely generous dogs and are at their happiest when everyone else is happy, including other teaching dogs. What is amazing, considering that they can be of the same breed and within the same pack, is that they function so very differently to the mentor and minder."

She scratched an itch on the side of her head before continuing, "The nanny dog not only relaxes a pack dog that is uncomfortable or being a tad antisocial but can expand that to helping relax a mentor or minder belonging to the same group."

"Mentors rarely get overly stressed in teaching situations but minders often take their role quite seriously and consequently can become tense when working. If a nanny dog sees another teaching dog, most likely a minder, showing stress the nanny will consciously use her body language to reduce the tension all around. That's why the nanny dog has been called the clown dog by some. Not in the sense of clowning around but through offering happiness to her fellow group members. It's fair to say that of all the teaching dogs the nanny dog is most likely to be happy in most situations."

Angela had clearly come to the end of what she had wanted to explain to me and for a couple of minutes the two of us remained quiet, me turning over the revelations of these aspects of dogs. Pharaoh was still sleeping on the settee.

I knew that I was in one of those rare moments, that of fully and comprehensively embracing an aspect of my life. Forevermore, a dog would not be some cute, cuddly pet but the modern, living embodiment of a species that not only has been with humankind for, literally, thousands of years, but has been instrumental in our human development for the past ten or fifteen thousand years, most probably many more years before that.

"Angela, I'm practically speechless and, trust me, that doesn't happen too often." There was a wry smile on my face that seemed to connect with her.

With the corners of Angela's mouth turned up in perfect harmony with my mood, she said, "I'm so pleased. Despite having seen hundreds of dog owners over my years, I was always puzzled by how few were motivated to understand, thoroughly, what makes the dog the animal that it is."

Pharaoh must have sensed some ending coming along as he shuffled up from his prone position on the settee cushion to sitting on his haunches. He was looking alertly towards Angela.

She continued, "So let's call it a day at this point. But I'll tell you what I think your plan should be."

She stood up, stretched her arms, and stifled a yawn with her right hand.

"Whoops, apologies, don't know where that came from. Been talking too much, I suspect."

Going on to say, "For a few weeks, why don't you bring Pharaoh up here once a week, twice a week if you can make it, and we'll reinforce the owner-dog relationship between the two of you. It will also give me a chance to get to know Pharaoh better, see how he reacts to some of the poor souls that I see here."

Adding, "But I have to say that there is very little doubt in my mind that Pharaoh is a beautiful and fantastic example of a minder teaching dog."

I turned that over in my mind for a few moments. "Angela, you need to tell me what the cost of his training would be."

"Well, normally," she replied, "I charge fifteen pounds for a training lesson. But in this instance, let's just run an open account for a while. Because, if you are happy for Pharaoh to be a teaching dog in helping sort out the dysfunctional dogs that come to me, then I would be paying you. Won't be a lot, I'm here to tell you, but it's all grist to the mill isn't it."

I went on to ask. "Angela, what's your view about walking Pharaoh in public places, such as Totnes High Street, for example? I just want to avoid any conflict between Pharaoh and another dog, or more importantly, another person."

"Good point, Paul. Of all the teaching dogs, the minder is the one dog that can make instant intuitive judgments about other dogs and of other people. It's utterly beyond us humans to be in mental harmony with both the speed of a minder dog's judgmental process and what that dog has instinctively cottoned on to. So rather than be less than perfectly relaxed when you are out and about with Pharaoh, get him comfortable in wearing a full muzzle. Full muzzles don't bother dogs once they associate wearing a muzzle with being out in interesting places. Don't leave it on Pharaoh at home or in the car, just put it on when you are going to be amongst people and dogs where there might be the slightest chance of aggravation."

Adding, "Mole Valley Farmers over at their store near Newton Abbot have a good selection."

I was a very, very happy man now.

"Oh, hang on a moment. Let me get you a copy of that article about the history of the family dog, that article by Dr. George Johnson."

A few minutes later I was swinging the car out of Angela's yard and starting the return journey to Harberton.

Finding the source of the River Dart would have to wait once again.

Pharaoh checking out Cleo, 2012.

What a magnificent animal. Pharaoh in the snows of Arizona, 2012.

Jean reading one afternoon in July 2011 with Sweeny on her lap and Hazel asleep by her feet.

Part Two:

Looking at Today

Chapter 6

AN EXPLANATION

After I had completed the first draft of the book I took a deep breath and gave it to a number of wonderful friends who had volunteered to read it. Without exception, the feedback was that Parts Two and Three didn't seem to match the title of the book. I fully understood where they were coming from but, nevertheless, that left me with a challenge. Let me explain.

In the Introduction to this book, I offered the opinion that many people, even with a minimum of awareness about the world that we all live in, are deeply worried about the future. Adding that I had the sense that the certainty of past times had gone, that the once-trusted models of society were failing. What on earth does this have to do with dogs?

The proposition behind the book is that modern society has lost sight of the principles, the values if you like, of what truly matters. That if we don't accept the corrosive nature of many aspects of modern life then humankind is facing a very challenging future. That's without factoring in that since 1987 the world has been increasing in total population by a billion people every twelve years (an increase in population of more than *22,800 a day*).

I was inspired to write the book for two reasons. First, when it comes to my time to die I want to know that my grandson is fully aware of how badly my generation has let the planet down. Thus this book is one tiny element in what I see as a growing awareness of the need for change. Second, change can only take place if there is a clear idea of what those changes have to be. This book argues that the qualities that we see in our longest and closest animal companion, the dog, are so familiar to millions of dog lovers that they represent a perfect emblem, a golden living example, of the values that we humans need to adopt if we are to bequeath a sustainable future to my grandson, and all the other millions of grandchildren alive today.

Now I would be the first to stick my arm up and admit the subjective and personal nature of what I have just written. Recognising that one person's sense of uncertainty is another person's sense of a hopeful future. There is also little doubt in my mind that the last thing you want to read is a catalogue of all the things that have the power to worry you and me. The daily news bulletins offer a very effective service in that regard!

So, please, as you read the chapters that comprise Parts Two and Three, do bear in mind that I am underpinning the core premise of this book. That premise that is articulated in detail in Part Four: the qualities that we need to learn from our dogs.

Chapter 7

THIS TWENTY-FIRST CENTURY

Bad news sells! Bad news also causes stress and worry. In my previous explanation, I explained that the last thing you want is a catalogue of all the things that have that power to cause you stress and worry. However, I do see three fundamental aspects of this new century that have their roots in that loss of principles that I referred to in the previous chapter. They are

1. the global financial system,
2. the potential for social disorder, and
3. the process of government.

Because they are at the heart of how the coming years will pan out.

The first aspect, our global financial system, was selected because it underpins all our lives in so many ways. When I was living in southwest England I was a client of Kauders Portfolio Services.[15] The founder of the company, David Kauders, published[16] a book, *The Greatest Crash*, in 2011. It was an obvious read for me at that time and I still have the book on my shelves here in Oregon.

David explained that whether we like it or not, our lives are inextricably caught up in the twin dependencies of the global financial system: credit and debt. As he wrote in his opening chapter:

> Households can barely afford their existing debts, let alone take on more. Since households now prefer not to borrow, indeed some even choose to pay back debt, it follows that those who have already borrowed, as a group, can no longer contribute to economic expansion.
>
> People can be divided into borrowers and savers. With existing borrowers unable to afford or unwilling to take on extra debt, can new borrowers be found instead? Those who do not need to borrow are unlikely to volunteer. Except for the young wishing to buy houses, facing the reality that house prices are beyond their pockets, where are the new borrowers?
>
> Businesses are also under pressure. There has been an inadequate recovery from recession, business prospects are poor as households cut back their spending. Lack of bank lending is a symptom rather than a cause, for if existing businesses were to be given more credit, they would probably be unlikely to find profitable growth opportunities in a world of austerity.

Later on in the book David describes this as "the financial system limit". In other words, the period of growth and expansion, especially of financial and economic expansion, has come to an end in a structural sense. This was his perspective from 2011.

Recently, I chose to reread *The Greatest Crash*. What struck me forcibly, reading the book again some four years later on, was how visible this "system limit" appeared in the world today. Everywhere there are signs that the era of growth has come to an end. Many countries are now indebted to a point that reinforces the proposition of there being a financial system limit. The United States is greatly in debt[17] but the only thing mitigating that situation, for the time being anyway, is that the American dollar is the quasi dominant global currency.

The changing nature of the global population is also reinforcing the fact that this is the end of a long period of growth. Even without embracing the question of how much longer we can increase the number of people living on a finite planet, the demographics spell out a greater-than-even chance of a decline in consumption and economic activity. Simply because in all regions of the planet, except for India where there is still a growing youth element in the country, people are ageing. To state the obvious, ageing persons do not consume as much as middle-aged and younger persons.

Thus, the world's economy that is just around the corner is certainly going to be very different to what it has been in the past. It is not being widely discussed. Worse than that, there is a widespread assumption adopted by many governments that a return to the "normal" economic growth of previous times is a given. Many do not share that assumption.

The second aspect that isn't being spoken about is the potential for massive, widespread social disorder. All summed up in just three words: greed, inequality, and poverty. Just three words that metaphorically appear to me like a round, wooden lid hiding a very deep, dark well. That lifting this particular lid, the metaphorical one, exposes an almost endless drop into the depths of where our society appears to have fallen.

Even the slightest raising of awareness of where this modern global world is heading is scary. I have in mind the author Thomas Piketty who warned[18] that, "the inequality gap is toxic, dangerous." Then there was the news in 2015[19] that, "Billionaires control the vast majority of the world's wealth, 67 billionaires already own half the world's assets; by 2100 we'll have 11 trillionaires, while American worker income has stagnated for a generation."

The third and final aspect that isn't being widely discussed is the process of government. Not from the viewpoint of "left" or "right", Labour or Conservative, Democratic or Republican (insert the labels appropriate to your own country), that is being discussed *ad nauseum*, but from the viewpoint of *good* government. It might be a terrible generalisation but it is still a fair criticism to say that many peoples of many countries have lost faith in their governments.

There appears to be a chronic absence of open debate about the need for good government, what that good government would look like, and how do societies bring it about.

If we were a dog pack, then our leader, our female mentor dog, would have moved us all to a new, pristine territory!

Phoebe, Lilly and Paloma sharing a bed, 2012.
Phoebe and Lilly have since died.

This is what living with dogs is all about. Jean relaxing on the bed with Hazel, Sweeney and Dhalia.

Chapter 8

BEHAVIOURS AND RELATIONSHIPS

"It is all to do with relationships." I heard this many years before the idea of writing this book came to me. Heard it from J one day when he was speaking of what makes for happy people in all walks of life. It's one of those remarks that initially comes over as such an obvious statement, akin to water being wet or the night being dark, that it is easy to miss the incredible depth of meaning behind those seven words.

Humans are fascinating. Every aspect of who we are can be seen in our relationships. How we relate to people around us, whether it be a thirty-second exchange with a stranger or a long natter with family and friends or partners whom we have known for decades. The core relationship, as in the relationship that drives our behaviours with other people, is the relationship that we have with ourselves. That core relationship being rooted, when we were young people, in our relationship experiences with the adults around us. For how we experienced being loved as a young person underpins that well-known saying, "You can't love another, if you can't love yourself."

Every aspect of how we live on this planet, how we care for everything in our domain in the broadest meaning of the word, comes down to relationships. If we care for ourselves, then we are gentle on our minds and bodies. If we love our partner and our family, then most of the time we think of their needs before our own. If we care for our friends and close associates, then we try to see their world through their eyes. That last thought brings to mind the saying: "Listen with intent to understand rather than with intent to respond."

But it doesn't stop there.

If we care for our planet, for the natural world and the wildlife, then we will do everything we can to actively protect the environment. If we care for our domestic animals, then we do everything to make their relatively short lives happy and content. As so many thousands and thousands do for their adorable dogs.

Everyone understands that in this new century we are not caring for our planet as we should be. It would be easy to condemn our drive for progress and an insatiable self-centredness as root causes of this deficiency. But it's not the case, certainly not the whole case.

The heart of the issue is clear to me. It is this. How we have developed is the result of human behaviour. All of us behave in many ways that are damaging, in varying degrees, to both the survival of our species and countless other animal species. It is likely that these behaviours are little changed over thousands of years.

But 2,000 years ago, the global population of humans was only 300 million persons[20]. It took 1,200 years for that global population to become, in 1800, 1 billion persons. The growth rate of global population is slowing[21] but nevertheless it is trending to a billion additional persons every decade.

Combine man's behaviours that are rooted in times long ago with this growth in population and we have the present situation. A totally unsustainable situation for one basic and fundamental reason. We all live on a finite planet.

The only viable solution is to amend our behaviours. To tap into the power of integrity, of self-awareness, and of mindfulness, and to change our game. In other words, to examine the fundamental relationship that we have with ourselves, the relationships that we have with one another and, above all, with the planet upon which we all depend. We need a level of consciousness that will empower change. Better than that, we need a level of consciousness *and integrity* that will empower change. And time is not on our side.

Pharaoh getting to know new arrival Oliver, June 2014.

Part Three:

Looking Forwards

Ruby relaxing at home, February 2014.

Chapter 9

THE DOOMSDAY CLOCK

"I can predict anything, except those things concerning the future!"

It's such a silly little saying, almost of schoolboy standards of humour. Yet I hear an underlying message of wisdom in those ten words. A message that connects strongly to another favourite saying of mine: "Never underestimate the power of unanticipated consequences."

Despite the obvious and rational assessment that making any form of reliable plan for the future is almost a contradiction in terms (apart from death and taxes) nonetheless we still invest in predicting the future in many ways. A common process is to identify key measures, or boundaries, that if crossed would herald a much more uncertain, as in risky, future.

Thus, since the 1990s we have had the warning that an average global surface temperature increase of more than 2 °C (3.6 °F) over the pre-industrial average runs the risk of dangerous, unpredictable climate change consequences. There is also concern that ocean acidification, as in a declining pH of the Earth's oceans, caused by the uptake of carbon dioxide (CO_2) from the atmosphere, could have a range of possible harmful consequences, especially involving that aspect of the food chain that is connected with the oceans.

Just a few years after I was born there was sufficient concern about the risk of a "cold war" that the Doomsday Clock came into existence. The origin of the clock was linked to an international group of researchers, called the Chicago Atomic Scientists, who had participated in the Manhattan Project[22]. Ergo, the world in 1947 became aware of the Doomsday Clock and the symbolic "minutes to midnight" likelihood of a global catastrophe. Since 1947 the setting of the minute hand has been maintained by the members of the Science and Security Board of the *Bulletin of the Atomic Scientists*. As that bulletin explains in the Overview[23]:

> The Doomsday Clock is an internationally recognized design that conveys how close we are to destroying our civilization with dangerous technologies of our own making. First and foremost among these are nuclear weapons, but the dangers include climate-changing technologies, emerging biotechnologies, and cybertechnology that could inflict irrevocable harm, whether by intention, miscalculation, or by accident, to our way of life and to the planet.

In researching more into this, I learnt that, "the Science and Security Board are advised by the Governing Board and the Board of Sponsors, that Governing Board including no less than eighteen Nobel Laureates." In other words, give or take some seventy years after the Doomsday Clock came into being, it still represents one of the better, as in more calmer, assessments of the future as of today.

Indeed, at the time of writing this book the most recent officially announced setting, as in three minutes to midnight (11:57 pm), was made on January 22, 2015. "The members of the Board were concerned about climate change, the modernisation of nuclear weapons in the United States and Russia, and the problem of nuclear waste."

As a matter of historical interest, the closest setting to midnight was just two minutes, set in 1953, and remained at two minutes until 1960, when it was reset to seven minutes to midnight[24].

Thus when it comes to a reasonable, considered assessment of the future for humankind and our dogs, and all other animals, then the Doomsday Clock is as good a symbol as one could wish for. But it is just a symbol. A symbol that could be interpreted as a cry for hope and change that I expand on in the next two chapters.

Chapter 10

THE SHARING OF HOPE

"The night is always darkest just before the dawn." This wonderful, famous saying rings out a single word for me. Rings out the single word: Hope.

Hope. What a short, little word to carry on its back, so to speak, so much promise, so many reasons for remaining resolute in one's ambition, for staying firm, for staying with it, whatever "it" might mean at any particular time for any particular person. Yes, for countless times, for countless numbers of people, the sharing of hope has been at the heart of embracing a better future. That hope underpins the belief of a "new dawn".

Now those fine, stirring words fail to highlight the many pragmatic ways in which communities of people demonstrate a commitment to hope. Here is an example of that pragmatism from my local county here in southern Oregon: Josephine County. Recently there was a county proposal for a property tax to be applied to homeowners to raise revenues to be used for increased public safety. The proposal failed. In 2014 there was a county proposal for increased financial support for county animal shelters. That proposal passed.

What that says to me is that the people of Josephine County have more hope in the outcome of better treatment for lost and rejected animals, including large numbers of dogs, than they do in better public safety for humans. In support of that premise, the staff at our nearest animal shelter, the Rogue Valley Humane Society, is reporting an increased awareness of the plight of animals in their care, of which dogs make up a sizeable proportion, and more animals going to loving homes.

It is almost as though people are reacting to the present global uncertainty by turning to their animals, to their dogs and cats.

Moving on to a rather more professional assessment, let me introduce Scott Barry Kaufman[25] the scientific director of the Imagination Institute in the Positive Psychology Center at the University of Pennsylvania. In December of 2011, he published an essay in the journal *Psychology Today*[26]. The essay was titled: "The Will and Ways of Hope".

The essence of Dr. Kaufman's proposition was that hope was generally undervalued both in society and, more specifically, within the field of psychology. In other words, while skills and talent will help you achieve whatever you aim for, it is hope that is the "vehicle" that delivers you to where you wish to be. In Kaufman's words, "Put simply: hope involves the will to get there, and different ways to get there."

When I read Kaufman's essay, I heard him saying that hope was not some wishy-washy emotion, some vaguely defined way of giving oneself a "good talking to" but instead it is "a dynamic cognitive motivational system", to use his words, and "cognitive" being the key element. That people who have clear goals in their lives, especially goals that demand learning, that in themselves are related positively to success, then those goals are more likely to be achieved if they are dependent on hope. Kaufman took the stance that awareness, or cognition, leads the emotions, not the other way around.

I found myself quietly wondering about where hope fitted in with other psychological attitudes, such as optimism and self-confidence. I didn't have to wonder for long because towards the end of the professor's essay he specifically addresses that, essentially saying that people with hope will most likely also have the willpower and the strategic clarity to achieve their specific goals.

For the first time in many decades of reflection and introspection into the aspects of success, motivation, and achievement, I saw hope in a vivid two-dimensional manner: the will and the ways. I had gained a clarity about hope as the power of bringing about change. The clarity that we really could emulate the wonderful ways our dogs behave and that would fuel the changes we so badly need.

I cannot better move on to the next chapter on change than by quoting Aristotle, "Hope is a waking dream."

Chapter 11

ON CHANGE

"They didn't bring us here to change the past" is from the film *Interstellar* that was drawing in the crowds when I was up to my neck in the first draft of this book. Jean and I had taken an afternoon off, together with neighbours Dordie and Bill, to go and watch the film. It was the middle of November 2014.

That spoken line really jumped out at me from the screen. Why? Simply because when it comes to making deep, fundamental changes in who we are, how we see ourselves and, flowing from that, how we behave, or better put, *how we wish to behave*, we have to change the past.

Clearly, we can't change the past in any real sense. But what we can change is our *understanding* of our past and how it made us the person we are now. That understanding of the "old you" is paramount if we are to set out along any journey of personal change. Anyone who has attempted a change in their behaviour, from a new year's resolution to a significant need to change, will appreciate the difficulty of achieving a *lasting* change in their behaviour. Changing our behaviour is rarely simple, straightforward or even, surprisingly, logical. Very often it requires a major commitment of our time and effort. Perhaps most important of all, that commitment to change is an emotional commitment that, as the previous chapter explained, is built around hope.

In Chapter 7, I implied that many aspects of these present times were unsustainable and that without change, the future was dark and uncertain. One explanation of why these feel like unsettling times, according to Paul Mason writing in *The Guardian* newspaper in July 2015, is that we are living through the transition between two eras; the transition from the era of capitalism to a new post-capitalist era. Mason, author of the book *Postcapitalism*[27] wrote in that article[28]: "As with the end of feudalism 500 years ago, capitalism's replacement by postcapitalism will be accelerated by external shocks and shaped by the emergence of a new kind of human being. And it has started." Mason's words stirred an ancient memory in me of something I recall reading in an Eckhart Tolle book, *The Power of Now*[29], about the power of turmoil and stress in bringing about enlightenment.

If Mason is correct then these uncertain times are part of the necessary transition to the new era ahead. That the next era, with new ways of sharing, learning, and working, will be different beyond comprehension to those of us living in today's world. Perhaps better expressed as those of us living in today's world who are the wrong side of 50!

The theme of this book, that the human-dog bond offers the most elegant and familiar means of learning the values and behaviours that we need for a sustainable future, is, by implication, about change. Change is always unsettling and brings out our need for safety. This psychological need for safety coupled with our hope for a better future is all wrapped up in the changes that are taking place now, evidenced by a whole range of actions and reactions across the world, including our changing relationships with our dogs. In 2013, *The Oregonian* newspaper reported that, "Pet ownership among single people has increased by nearly 17 percent, from 46.9 percent in 2006 to 54.7 percent in 2011, according to a recent survey by the American Veterinary Medical Association." Later revealing in that same article, "The study also indicates that singles are more likely to think of pets as family members, rather than companions or property."

Thus this growing sense of insecurity might be the cause of why many are increasingly relying on families and local communities, rather than our governments, to protect us, and increasingly relying on our dogs, a real throwback to very ancient times. In the Foreword to this book, Jim Goodbrod speaks of dogs demonstrating, "a simple and uncorrupted approach to life from which we all could benefit." Part Four goes into the details of that simple and uncorrupted approach to life.

Somewhere, now long ago forgotten, I came across this quotation from the notable management writer Simon Caulkin: "It's all the product of human conduct." Caulkin puts his finger on the truth about humankind's future. Everything to do with us is the product of our conduct. It is *only* human conduct that will find that sustainable, balanced relationship with one another and, critically, with the planet on which all our futures depend.

When writing that last paragraph, I was aware of another idea floating to the surface of my mind. Something that linked this chapter on change to the previous chapter on hope and the idea of us having faith in humankind. Faith that not only we *can* change our relationship with ourselves, with our communities and, above all, with our planet, but that *we will*. Faith that we, as in all humankind, will embrace the many beautiful qualities of the animal that is so special to so many millions of us: our dogs. Not just embrace those qualities but pin our future on the premise that adopting the qualities of integrity, love, trust, honesty, openness and more, qualities that we see daily in our closest animal companions, is our potential salvation.

My final perspective about change has to do with our human genders: men and women. Namely, that the evolution of humankind, including critically our ability to cooperate flexibly in large numbers, is rooted in the gender differences between man and woman. I opine that until, say one hundred years ago, give or take, cooperation among large numbers of us humans was critically important in so many of life's areas: health, science, medicine, physics, exploration, outer space, and more. (And whether one likes it or not: wars.) My argument is that it is predominantly men who have been the "shakers and movers" in these areas. Of course not exclusively, far from it, just me saying that so many advances in society in the past have more likely been led by men.

But I contend that these present times call for a different type of man (and woman). A human who is less the rational thinker, wanting to set the pace, and more a human capable of expressing fears, exploring feelings, defining fear of failure, and more. In August 2015, Raúl Ilargi Meijer wrote a post on his blog *The Automatic Earth*. His post was called "Power and Compassion".[30] It contained the following:

> For a society to succeed, before and beyond any economic and political features are defined, it must be based solidly on moral values, a moral compass, compassion, humanity and simple decency among its members. And those should never be defined by economists or lawyers or politicians, but by the people themselves. A social contract needs to be set up by everyone involved, and with everyone's consent. Or it won't last.

For that social contract to be successfully set up by everyone must, of course, include men and women. And that requires all people to find safe ways to get in touch with their feelings, to tap into their emotional intelligence, using positive psychology to listen to their feelings and know the truth of what they and their loved ones need to guarantee a better future. What they need in terms of emotional and behavioural change. Of course, more women understand what is meant by social contracts as they are predominantly caring, social humans. But it is good to remind us all that this is an essential part of what we must do.

Finally, dogs neither make nor comprehend social contracts but they demonstrate the *simple and uncorrupted approach to life* that we humans surely need to adopt. It's never too late to change.

Hazel with her wonderfully expressive eyes.

Part Four:

Learning from Dogs

Chapter 12

SETTING THE SCENE

You will recall that Chapter 5 was about me discovering that my German Shepherd dog, Pharaoh, was a teaching dog of the type known as a minder. It was an account of how in November 2003, when living in England, I was made aware of the rich and diverse attributes of the species *Canis lupus familiaris*, thanks to Angela Stockdale.

It is not central to the purpose of this book, about what we must learn from our dogs, to elaborate on the journey that led me from living in the county of Devon, southwest England to living now in Josephine County, a rural part of southern Oregon, in the United States. Still with dear old Pharaoh, but also with nine other examples of the species *Canis lupus familiaris*. Indeed, since 2008 I have been living with nine and at times up to fourteen dogs in my life. What I have learnt from having so many dogs around me makes Part Four of this book possible.

Chapter 13

INTEGRITY

It is a Friday morning in June in the year 2007. I am sitting with J at his place in South Devon with Pharaoh sleeping soundly on the beige carpet behind my chair. I didn't know it at the time but it was to become one of those rare moments when we gain an awareness of life that forever changes how we view the world, both the world within and the world without.

"Paul, I know there's more for me to listen to and I sense that we have established a relationship in which you feel safe to reveal your feelings. However, today I want to talk about consciousness. Because I would like to give you an awareness of this aspect of what we might describe under the overall heading of mindfulness."

I sat quietly fascinated by what was a new area for me.

"During the years that I have been a psychotherapist, I've seen an amazing range of personalities, probably explored every human emotion known. In a sense, explored the consciousness of a person. But what is clear to me now is that one can distil those different personalities and emotions into two broad camps: those who embrace truth and those who do not."

J paused, sensing correctly that I was uncertain as to what to make of this. I made it clear that I wanted him to continue.

"Yes, fundamentally, there are people who deny the truth about themselves, who actively resist that pathway of better self-awareness, and then there are those people who want to know the truth of who they are and actively seek it out when the opportunity arises. The former group could be described as false, lacking in integrity and unsupportive of life, while the latter group are diametrically opposite: truthful, behaving with integrity and supportive of life."

It was then that J lit a fire inside me that is still burning bright to this day. For he paused, quietly looking at Pharaoh sleeping so soundly on the carpet, and went on to add, "And when I look at dogs, I have no question that they have a consciousness that is predominantly truthful: that they are creatures of integrity and supportive of life."

That brought me immediately to the edge of my seat, literally, with the suddenness of my reaction causing Pharaoh to open his eyes and lift up his head. I knew in that instant that something very profound had just occurred. I slipped out of my chair, got down on my hands and knees and gave Pharaoh the most loving hug of his life. Dogs are creatures of integrity. Wow!

Later, when driving home, I couldn't take my mind off the idea that dogs were creatures of integrity. What were those other values that J had mentioned? It came to me in a moment: truthful and supportive of life. Dogs have a consciousness that is truthful, that they are creatures of integrity and supportive of life: what a remarkable perception of our long-time companions.

I had no doubt that all nature's animals could be judged in the same manner but what made it such an incredibly powerful concept, in terms of dogs, was the unique relationship between dogs and humans, a relationship that went back for thousands upon thousands of years. I realised that despite me knowing I would never have worked it out on my own, J's revelation about dogs being creatures of integrity was so utterly and profoundly obvious.

As I made myself my usual light lunch of a couple of peanut butter sandwiches and some fruit and then sat enjoying a mug of hot tea, I still couldn't take my mind off what J had revealed: dogs are examples of integrity and truth. I then thought that the word "examples" was not the right word and just let my mind play with alternatives. Then up popped: Dogs are *beacons* of integrity and truth. Yes, that's it! Soon after, I recognised that what had just taken place was an incredible opening of my mind, an opening of my mind that didn't just embrace this aspect of dogs but extended to me thinking deeply about integrity for the first time in my life.

Considering that this chapter is titled "Integrity", so far all you have been presented with is a somewhat parochial account of how for the first time in my life the word "integrity" took on real meaning. That until that moment in 2007 the word had not had any extra significance for me over the thousands of other words in the English language. Time, therefore, to focus directly on integrity.

"If goodness is to win, it has to be smarter than the enemy."

That was a comment written on my blog some years ago, left by someone who writes their own blog under the nom-de-plume of Patrice Ayme[31]. It strikes me as beautifully relevant to these times, times where huge numbers of decent, law-abiding folk are concerned about the future. Simply because those sectors of society that have much control over all our lives do not subscribe to integrity, let alone giving it the highest political and commercial focus that would flow from seeing integrity as an "adherence to moral and ethical principles; soundness of moral character; honesty." To quote my American edition[32] of Roget's Thesaurus.

Let me borrow an old pilot's saying from the world of aviation: "If there's any doubt, there's no doubt!"

That embracing, cautious attitude is part of the reason why commercial air transport is one of the safest forms of transport in the world today. If you had the slightest doubt about the safety of a flight, you wouldn't board the aircraft. If you had the slightest doubt about the future for civilisation on this planet, likewise you would do something! Remember, that dry word civilisation means family, children, grandchildren, friends, and loved ones. The last thing you would do is to carry on as before!

The great challenge for this civilisation, for each and every one of us, is translating that sense of wanting to change into practical, effective behaviours. I sense, however, that this might be looking down the wrong end of the telescope. That it is not a case of learning to behave in myriad different ways but looking at one's life from a deeper, more fundamental perspective: living as a person of integrity. So perfectly expressed in the Zen Buddhist quote: "*Be master of mind rather than mastered by mind.*" Seeing integrity as the key foundation of everything we do. Even more fundamental than that. Seeing integrity as everything you and I are.

It makes no difference that society in general doesn't seem to value integrity in such a core manner. For what is society other than the aggregate of each and every one of us? If we all embrace living a life of integrity then society will reflect that.

Integrity equates to being truthful, to being honest. It doesn't mean being right all the time, of course not, but integrity does mean accepting responsibility for all our actions, for feeling remorse and apologising when we make mistakes. Integrity means learning, being reliable, and being a builder rather than a destroyer. It means being authentic. That authenticity is precisely and exactly what we see in our dogs

The starting point for what we must learn from our dogs is integrity.

Jean and Pody sharing a start to the day.

Chapter 14

LOVE

It is incredibly easy and yet so difficult to write about the love of a dog. Now if that isn't a paradox, then I don't know what is!

Let me try to explain this more rationally.

Dogs are so quick to show their love for a human. It could be the wag of a tail, the way a dog's eyes connect with our eyes, a gentle lean of a head against our legs. Or a curling up on our lap, a lick or two of our hands or our faces, and more, so much more. All of these ways make sense to us. For they are familiar to us humans from the point of view of how we show our love to our partner or to our children. Perhaps not as directly as licking their hands or faces, but you know what I am trying to say!

However, there are countless numbers of stories about dogs offering love to us humans that go way beyond anything that we could emotionally understand. Let me offer one that was published[33] on my blog in January 2011. It was the story of a Skye Terrier called Bobby.

> On 15 February 1858, in the City of Edinburgh in Scotland, a man named John Gray died of tuberculosis. Gray was better known as Auld

Jock and after his death he was buried in the Greyfriars kirkyard situated on Candlemaker Row in Edinburgh.

Bobby had belonged to John Gray, John having worked for the Edinburgh city police as a night watchman. The two of them, John and Bobby, had been virtually inseparable for the previous two years.

When it came to the funeral, Bobby led his master's funeral procession to the grave at Greyfriars cemetery and later, when this devoted Skye Terrier tried to stay by the graveside, he was sent away by the church caretaker. But Bobby returned and refused to leave whatever the weather conditions. Despite the best efforts of the keeper of the kirkyard, plus John's family and many local people, Bobby refused to be enticed away from the grave for any length of time and, as a result, he touched the hearts of all the local residents.

Although technically dogs were not allowed in the graveyard, people rallied round and built a shelter for Bobby and there he stayed guarding Auld Jock, his late master. Yes, there Bobby stayed, stayed for fourteen years, laying on the grave, leaving only for food.

To this day, close by Greyfriars kirkyard, there is a Bobby's Bar and outside the bar a cast-metal stature of Bobby on a plinth.

The love that a dog shows us is a form of unconditional love that is not unknown in our human world but is not common. I would vouch that few people have truly ever experienced unconditional love from another person or are even clear as to what it is. For although one might define unconditional love as affection without any limitations or love without any conditions, in other words a type of love that has no bounds and is unchanging, the reality of the love of one person towards another (whether it be a spouse, lover or child) is that there are potentially limits to that love. If there were none then divorce and separation would not exist. Now I'm not suggesting that a dog will offer unconditional love to a human who abuses that dog. But I am saying that dogs exhibit a loving loyalty that is practically unrecognised in our human world.

Let me turn to the world of novels. Authors, at times, make a distinction between unconditional love and conditional love. In that, conditional love is earned through conscious or unconscious behaviours that reach out to the lover. Whereas unconditional love is illustrated when love is given no matter what. Another way of looking at unconditional love is to see it as an expression of feelings *irrespective* of the will of the lover.

Yet there's another aspect of unconditional love that commonly applies between individuals and their dogs. That is that our love for our dog encompasses a desire for the dog to have the very best life in and around us. Take the example of a new puppy. The puppy is cute, playful, and the owner's heart swells with love for this adorable new family member. Then the puppy urinates on the floor. One does not stop loving the puppy but recognises the need to modify the puppy's behaviour through love and training rather than continue to experience behaviours that would be unacceptable in particular situations.

One of the things that frequently has moved me to tears is how quickly a dog, new to our home, has settled in and become certain that our love for that new member of our doggy family is solid.

In April 2015 Jean and I went to a local farmers' supplies store that we use on a regular basis for stuff, including dog food. Within the store the local town animal shelter had a number of dogs on show seeking homes for them. One of the dogs was a small five-year-old Chihuahua mix that we found irresistible and Pedy, for that was his name, became our latest love.

Within just twenty-four hours of being brought home, not only had he made friends with the rest of our dogs, he had found many places to sleep, including Jean's lap. Pedy had been brought to the animal shelter after having been found abandoned in a rural area about twelve miles from town. It was clear that for most of his life he had been a pet and the staff watching the various cages at the store thought that he had not been disposed of that long ago.

Pedy showed no scars from his recent traumatic experience and approached both Jean and me and all the other dogs from a place of unconditional love. He showed that such love has the power to attract love back, as it did from the other dogs and from my wife and me. It was a wonderful example of what can be gained through such a love.

But it is not the only example of a dog offering unconditional love, far from it.

In 2003, police in Warwickshire, England, opened a garden shed and found a whimpering, cowering dog. The dog had been locked in the shed and abandoned. It was dirty and malnourished, and had quite clearly been abused. In an act of kindness, the police took the dog, which was a female greyhound, to the Nuneaton Warwickshire Wildlife Sanctuary, which is run by Geoff Gruecock and known as a haven for animals abandoned, orphaned, or otherwise in need.

Geoff and the other sanctuary staff went to work with two aims: to restore the dog to full health and to win her trust. It took several weeks, but eventually both goals were achieved. They named her Jasmine and they started to think about finding her an adoptive home. Jasmine, however, had other ideas. No one quite remembers how it came about, but Jasmine started welcoming all animal arrivals at the sanctuary. It would not matter if it was a puppy, a fox cub, a rabbit, or any other lost or hurting animal. Jasmine would just peer into the box or cage and, when and where possible, deliver a welcoming lick.

Geoff relates one of the early incidents. "We had two puppies that had been abandoned by a nearby railway line. One was a Lakeland Terrier cross and another was a Jack Russell Doberman cross. They were tiny when they arrived at the centre, and Jasmine approached them and grabbed one by the scruff of the neck in her mouth and put him on the settee. Then she fetched the other one and sat down with them, cuddling them."

Geoff continued, "But she is like that with all our animals, even the rabbits. She takes all the stress out of them, and it helps them to not only feel close to her, but to settle into their new surroundings. She has done the same with the fox and badger cubs, she licks the rabbits and guinea pigs, and even lets the birds perch on the bridge of her nose."

Jasmine, the timid, abused, deserted waif, became the animal sanctuary's resident surrogate mother, a role for which she might have been born. The list of orphaned and abandoned youngsters she has cared for comprises five fox cubs, four badger cubs, fifteen chicks, eight guinea pigs, two stray puppies, fifteen rabbits, and one roe deer fawn. That fawn, named Tiny Bramble, eleven weeks old, was found semi-conscious in a field. Upon arrival at the sanctuary, Jasmine cuddled up to her to keep her warm, and then went into the full foster-mum role. Jasmine, the greyhound, showered Bramble, the roe deer, with affection.

Now offering love to the world in the unconditional manner of Pedy, Jasmine and countless other dogs might be a step too far for us humans but that doesn't negate the urgent need for us to adopt a truly loving approach to the world. Here's how that might be achieved.

"It's the little things that count is a famous truism" and one no better suited to the world of love. Little things that we can do in countless different ways throughout the day.

- Sharing a friendly word or a smile with a stranger.
- Dropping a coin or two into a homeless person's hands or, better still, a loaf of bread or a chocolate bar.
- Being courteous on the road.
- Holding a door open for someone at your nearby store.
- Showing patience in a potentially frustrating situation.

Continuing, perhaps, with such little things as never forgetting that we have two ears and one mouth and should use them in that proportion, or being more attentive when a loved one is speaking with us. Actively embracing periods of quiet contemplation, or understanding that the world will not come to an end if the television or smartphone is turned off for a day.

The list of loving actions we can engage in is endless but the actions will not become second nature unless we hang on, likewise endlessly, to why we need this love. For this world of ours so desperately needs a new start and that new start must come from a loving attitude to one another, to all the plants and animals, and to the blue planet that sustains us.

We need our hearts to open sufficiently to tell our heads about the world of love. Or in the words of Nelson Mandela,

> No one is born hating another person. People must learn to hate, and if they can learn to hate, they can be taught to love, for love comes more naturally to the human heart than its opposite.

No-one would argue with that!

Little Pedy, April 2015.

Hazel demonstrating her love for the author.

Chapter 15

TRUST

"The absolute certainty in the trustworthiness of another"[34] is not a bad definition of trust and one that dogs seem to embrace so much more easily than humans do.

Invariably, a dog trusts a new human in its life very soon after that first meeting. The speed at which a dog trusts a new person implies that there are many non-verbal cues that we give out and that the dog, almost instinctively, reads our body language and, when appropriate, feels safe.

It's easy for me to understand the trust in a dog when I look at Pharaoh for he has been part of my life since a few weeks after he was born in South Devon, England, on June 3, 2003. So my experience of Pharaoh doesn't offer any insight as to how a dog that has been cruelly treated by other people learns to trust again. I had to wait until 2010 to learn that lesson.

That was the year Jean and I decided to move to the States and be married. I met Jean in December 2007 and moved from England, with Pharaoh, to be with her in Mexico in 2008. Jean had been born in London, as I had, but circumstances had led her to living in Mexico for a number of years and she was well-known for rescuing abandoned dogs. At the time that we met, Jean had sixteen dogs, all of them rescues off the streets in and around San Carlos.

A couple of weeks before we were due permanently to leave San Carlos with all our animals and belongings and journey the 513 miles to Payson, in Arizona, Jean went outside to the front of the house to find a very lost and disorientated black dog alone on the dusty street. The dog was a female and because the dog's teats were still somewhat extended Jean surmised that not that long ago the dog had given birth to puppies that had recently been weaned. There was no doubt that the dog had been abandoned in the street. A not uncommon happening because many of the local Mexicans knew of Jean's rescues throughout the years and when they wanted to abandon a dog, frequently it was done outside her house. The poor people of San Carlos sometimes resorted to selling puppies for a few pesos and then casting the mother dog adrift.

Without delay the dog was taken in and we named her Hazel. Now, as humans we can't even start to imagine the emotional and psychological damage that a mother dog would incur from having had all her puppies removed from her so soon after the pups had been born. It's very unlikely that we could imagine the emotional damage a human mother would receive from the catastrophic, forced removal of her young baby.

The one thing that it would be reasonable to assume was that Hazel would initially be a bit wary of "the species that walks on two legs". Especially two people who were heavily distracted with the task of moving to another country in fewer than twenty days.

Once Hazel had been fed and watered by Jean, her coat inspected for ticks and generally checked all over, the next step was to introduce her to some of the other dogs. It all went very smoothly. Then later in the evening, the evening of the same day that we took Hazel in, when we were sitting down after our evening meal, Hazel came over to the settee and looked up into my eyes. In a way that couldn't be put clearly into words, I sensed a lost soul in Hazel's eyes and her desperate need for some loving. Those thoughts that I was having, that she was a lost soul desperate for love, were, I hoped, available for Hazel to read in my eyes.

Then, miracle of miracles, after a pause of a half-minute or so, our eyes still locked together, Hazel climbed up next to me on the settee and carefully and cautiously settled her head and front paws across my lap. I caressed and stroked her for much of the rest of the evening. When later on Jean and I went to bed, Hazel jumped up and went to sleep on the bed staying alongside my legs for the whole of the night setting a pattern that has continued to this very day. How did Hazel know to trust me? Only Hazel knows the answer. But ever since that day in February 2010, the bond between Hazel and me has been perfect. In fact, as I write these words, Hazel is asleep on the rug just behind my chair.

I closed Chapter 1 with these words, "All we can do is to dream about how that first coming together between human and wolf might have happened." It seems highly appropriate in a chapter exploring the quality and value of trust in dogs to relate a true story about trust between a man and a wolf, the genetic ancestor of the dog.

This story of an amazing relationship between a wild wolf and a man was told to me by DR when I was first living with Jean in San Carlos, Mexico; the story is about Tim Woods, a brother to DR.

Ex-wild wolf Luna resting close to Tim Woods.
Permission to use photograph granted by DR.

It involved the encounter of Tim, sleeping rough in a shack with his dog on Mingus Mountain, with a fully grown female grey wolf. By way of background, Mingus Mountain is located in the US state of Arizona in the Black Hills range and within the Prescott National Forest, approximately midway between Cottonwood and Prescott.

DR and brother Tim were part of a large family comprised of a total of seven sons and two daughters. Tim had a twin brother, Tom, and DR knew from an early age that Tim and Tom were different.

As DR explained:

> Tim was much more enlightened than the rest of us. I remember that Tim and Tom, as

> twin brothers, could feel each other in almost a mystical manner. I witnessed Tom grabbing his hand in pain when Tim stuck the point of his knife into his (Tim's) palm. Stuff like that. Tim just saw more of life than most other people.

The incident involving the wolf was when Tim was in his late 40s and, as previously mentioned, was living with his dog in a rough shack on Mingus Mountain. The shack was simply a plywood shelter with an old couch and a few blankets for the cold nights. The dog was a companion, a guard dog, and a means of keeping Tim in food, for the dog was a great hunter. But Tim was no stranger to living in the wild. For DR explained that Tim was ex-US Army and a great horseman and that there was a time when Tim was up in the Superstition Mountains, also in Arizona, sleeping rough and riding during the day. At night Tim would get the horse to lie down and he would sleep with his back next to the horse for warmth.

Back to the Mingus Mountain story, in DR's words:

> Anyway, Tim was up on Mingus Mountain using an old disk from an agricultural harrow as both a cook-pan and plate. After he had finished eating, Tim would leave his plate outside his shack. It would be left out in the open over night.
>
> Tim gradually became aware that a creature was coming by and licking the plate clean and so Tim started to leave scraps of food on the plate. Then one night, Tim was awoken to the noise of the owner

> of the tongue and saw that it was a large, female grey wolf. The wolf became a regular visitor and Tim became sure that the wolf, now having been given the name Luna by Tim, was aware that she was being watched by a human.
>
> During many, many months Luna built up sufficient trust in Tim that eventually she would take food from Tim's outstretched hand. It was only now a matter of time before Luna started behaving more like a pet dog than the wild wolf that she was. From then on, Luna would stay the night with Tim and his dog, keeping watch over both of them.

DR also recalled[35]:

> I remember Tim being distraught because, without warning, Luna stopped coming by. Then a few months later, back she was. Tim never did know what lay behind her absence but guessed it might have been because she went off to have pups.

Unfortunately, this wonderful tale does have a sad ending. Namely, that in 2007 Tim was awakened to hear a pack of coyotes yelping and his dog missing, never to be found again. Then tragically, some six months later Tim contracted a gall bladder infection. Slowly it became worse and by the time Tim realised that it was sufficiently serious to require medical treatment, it was too late. Despite the best efforts of modern medicine Tim died on 25 June 2009. He was just 51 years old.

DR closed his tale with these words: "So if you are ever out on Mingus Mountain and hear the howl of a wolf, reflect that it could be poor Luna calling out for her very special man friend."

Now very few of us are ever going to be out on Mingus Mountain but possibly there is still a way that we can connect with Tim. For wherever you are in the world, if you ever hear the howl of a wolf, give yourself the space to disappear into your inner thoughts for a few precious moments and know that tens of thousands of years ago there was another "Tim" and another "Luna". Maybe sense that first step in the long journey of human and dog, of that mythical Tim cuddling up to that mythical Luna. Or, perhaps, the next time you gaze deeply into your dog's eyes, know how far the trust offered by that wolf has brought all of us and our dogs.

Our society only functions in a civilised manner when there is a predominance of trust around us. When we trust the sociopolitical foundations of our society, when we trust the legal processes, and when we trust that while greed and unfairness are never absent, they are kept well under control.

Having trust in the world around us is an intimate partner to having faith in that same world. For without trust there can be no faith, and without trust there can be no love.

Jean and little Sweeny and Cleo.

Chapter 16

COMMUNITY

"If your plan is for one year plant rice.
If your plan is for ten years plant trees.
If your plan is for one hundred years educate children."
– Confucius

What do we mean by community? Until a few hundred years ago, a community would have been people living as a cohesive group in the same geographical place recognising the same cultural values. However, as untold thousands of people later on settled in different places from their place of birth, not infrequently in different countries, then those enduring community values have become progressively weakened.

In this current century, there is no let up in the levels of migration both to and from one's country of birth. However, this author senses a rebirth of that notion of community that is very much a product of these modern times. For I am referring to the widespread use of online social media sites and, to a lesser extent, blogging sites. To put figures to that widespread use, Facebook, founded in February 2004, now has about 1.4 billion monthly users. Twitter, founded in March 2006, is now running with about 302 million monthly users. And the earliest of the three, LinkedIn, formed in December 2002 and launched in May 2003, now[36] has about 300 million users.

All of these social-media sites, and others beyond them, create links among like-minded people way beyond our direct friends and members of our family. It seems to be a valid proposition to say that we now have communities of like-minded people, subscribing to the same social rules totally disconnected from the physical, as in geographical, connections of the past. We now have virtual communities right around the world.

This is a wonderful aspect of modern communication technologies. Because the future for all of us on this planet depends on effective communities. Only such twenty-first-century style communities will be able to push back against the overwhelming forces of power, corruption, and control.

Now let's turn to dogs, for dogs offer many beautiful examples of the benefits of community. For the reason that their ancient genes, long before they became domesticated animals, still guide their behaviours. When dogs lived in the wild, their natural pack size was about fifty animals and there were just three dogs that had pack status: the mentor, minder and nanny dogs, as described in Chapter 5. As was explained in that chapter, all three dogs of status are born into their respective roles and their duties in their pack are instinctive. There was no such thing as competition for that role as all the other dogs in that natural pack grouping would be equal participants with no ambitions to be anything else.

Anyone who has had the privilege of living with a group of dogs will know beyond doubt that they develop a wonderful community strength. Let's reflect on the lessons being offered for us in this regard by our dogs.

Many now think that governments are less effective at understanding the needs of their peoples than in the past. I'm thinking how the top echelons in many of our societies are disconnected from the needs and aspirations of their peoples. The widespread sense that representative democracy, as a process, is broken. Another feature of this present world is that unless one is living in a genuinely rural part of a country, those old-style communities have more or less disappeared.

In an online essay[37] that I came across written by Erik D. Kennedy titled "On the Social Lives of Caveman", he offers some powerful arguments for re-embracing a community way of life.

> Human beings are no strangers to group living. Call it a family trait. Our closest animal relatives spend a good bulk of their time eating bugs off their friends' backs. While I'm overjoyed we're not social in that manner, I'm less pleased that we're not social more to that *degree*. In study after study, having and spending time with close friends is consistently correlated with happiness and well-bcing. And yet, the last few decades in America have seen a remarkable decline in many things associated with being in a tight-knit social circle, things like family and household size, club participation, and number of close friends. Conversely, we've seen an increase

> in things associated with being alone, TV, commutes, and the internet, for example.
>
> This trend is quite unhealthy. It's no surprise that humans are social animals but it may be surprising that we're *such* social animals that merely joining a club halves your chance of death in the next year, or that living in a close-knit town of three-generation homes can almost singlehandedly keep you safe from heart disease.

The phrase, "we're such social animals that merely joining a club halves your chance of death in the next year" should be in the forefront of everyone's mind who is over the age of sixty-five!

One doesn't need to reflect long about community living and the example of our wild dogs before an obvious question arises: If fifty dogs is the optimum number for a pack of dogs, is there an optimum number of people in the human equivalent of a pack? Erik Kennedy[38] addresses that question:

> So we'll spend more time with other people. Fine. But who should we spend our time with? What kind of groups should we hang out in? And how big of groups? The simple answer is: as long as you're pretty close to the people you're with, it hardly matters. Piles of research back up what is essentially obvious from everyday experience: that the more time you spend with people you trust, the better off you are. That's not to discourage actively meeting new people, but seeing as though close friends push us towards health and

> happiness better than strangers, there does appear to be a limit on the number of people you can have in your "tribe".
>
> And that number is about 150, says anthropologist Robin Dunbar, who achieved anthropologist fame by drawing a graph plotting primates' social group size as a function of their brain sizes. He inputted the average human brain size into his model, and lo and behold, the number 150 has been making a whirlwind tour of popular non-fiction books ever since. Beyond being the upper bound for both hunter-gatherer tribes and Paleolithic farming villages, it appears that everything from startup employee counts to online social networks show this number as a fairly consistent maximum for number of close social ties.

The term paleo-social is not one that is commonly used now but what it represents, the modern equivalent of a tribe, might be common all over the world. In fact, paleo-social lives might not only be common but an age-old wisdom in many parts of the globe. In essence, therefore, adopting such a community lifestyle is not without precedent and needs to be reinvigorated in the Western world.

Such reinvigoration might be easily accomplished through practical actions. These simple changes to our lifestyles can deliver great community returns, such as watching less television, living in and around bigger groups, sharing projects with neighbours, and dining with the same people more frequently. Then there is the most important act of always resolving any disputes that you have with friends. When it comes to our younger ones, then good common-sense actions would be to have your children spend more time with trusted adults and, in turn, you spending time with the kids of adults who trust you and, perhaps, mixing up age groups wherever possible.

For one thing is clear beyond doubt. Humans are not designed to be alone; isolation and loneliness are damaging to our mental health. One might go so far as to say that the rewards of sharing, of living a local community life, might just possibly be the difference between the failure and survival of the human race. Now that is something we need to learn from our dogs.

Chapter 17

PLAY

Can we learn from the way that dogs play? Not in the metaphorical sense but in a direct sense?

It might seem perfectly reasonable to imagine that the answer to that question is no. That we have nothing to learn, in a real and tangible manner, from the playing of dogs. However, that answer would be wrong.

On May 19, 2014, the *Washington Post* published[39] an article titled: "In Dogs' Play, Researchers See Honesty and Deceit, Perhaps Something Like Morality." It was written by David Grimm[40], the journalist and online news editor of *Science Magazine* and the author of the book[41] *Citizen Canine: Our Evolving Relationship with Cats and Dogs*. Significantly, Marc Bekoff[42], first briefly mentioned in Chapter 1, is referred to extensively within the article. I used the word significantly because Marc Bekoff is professor emeritus of ecology and evolutionary biology at the University of Colorado, Boulder, and someone who has written much on the behaviours of animals.

In May 2015, Bekoff published an online essay[43] on the website of *Psychology Today*. The essay, titled "Butts and

Noses: Secrets and Lessons from Dog Parks" is his analysis of research that he had undertaken into the behaviours of dogs in dog parks, an exploration to which the professor had clearly devoted many hours. As Bekoff writes in the opening paragraph (my bold added for emphasis):

> Dog parks are a fascinating recent and growing cultural phenomenon. Indeed, I go rather often to what I call my field sites, for that's what they are, to study play behavior and other aspects of dog behavior including urination and marking patterns, greeting patterns, social interactions including how and why dogs enter, become part of, and leave short-term and long-term groups, and social relationships. I also study human-dog interactions and **when I study how humans and dogs interact I also learn a lot about the humans.**

Later on the professor writes, referring to dogs:

> Because play is a foundation of fairness there is a good deal of cooperation among the players as they negotiate the ongoing interaction so that it remains playful. I think one can make a good case for their having a theory of mind. Nonetheless, we still need more data on this aspect of play as well.

Now in our human world, and most certainly when it comes to sports and competitive games, fairness is the foundation stone of the rules associated with that sport or game. Therefore, I was struck by Marc Bekoff's analysis that for dogs play is "a foundation of fairness", not the other way around as it is with us humans.

Thus the implication that studying how dogs play could eventually throw light on the evolution of our human emotions is more than just a fascinating idea. The inference seen by me, in both the *Washington Post* article by David Grimm and Professor Bekoff's essay, is that from understanding how dogs play comes the evolutionary understanding of our emotions and from there, in the words of Grimm, "how we came to build a civilisation based on laws and cooperation, empathy and altruism."

That's quite something to learn from our dogs.

Grimm's article also reveals that

> Bekoff wasn't the first scientist to become intrigued by the canine mind. Charles Darwin in the mid-1800s had postulated that canines were capable of abstract thought, morality and even language. (Darwin was inspired by his own mutts; he owned 13 of them during his life.) Dogs, he wrote, understand human words and respond with barks of eagerness, joy and despair. If that wasn't communication between the species, what was?

Like millions of other dog lovers, I know from strong personal experience that dogs have a great sensitivity to how I am behaving and feeling. Almost taking it for granted that when I yawn, the chances are that one of our dogs will yawn. Or believing, without any doubt, that dogs show empathy for humans; I can easily recall my Pharaoh licking tears from my face. What I didn't realise until reading the *Washington Post* article is that empathy is a rarely documented trait within the animal kingdom.

We know what our dogs are feeling from their behaviour and their vocal sounds. Know instinctively that when a dog nudges me awake in the early hours of the morning, it is because it needs to go outside for a "call of nature" and can't wait until the normal waking hour. In return, our dogs know what we are feeling from our behaviours, our body language, and our vocal sounds.

Back to the subject of play. Science is clearly suggesting that play, as demonstrated so beautifully by Bekoff's analysis of dogs and other animals, is incredibly important for our own species. That without play, we would have had an impossible task of interacting with the world around us. That our insight into our human emotions and the way we conduct ourselves, in a social context, flows down from that learning from our dogs.

Leaving one inescapable conclusion, one that so perfectly links to community: never stop playing. Never stop playing with other people and, above all, never stop playing with our dogs.

Chapter 18

SHARING

Here's a silly story that made me laugh when I first came across it.

> A man in a casino walks past three men and a dog playing poker.
> "Wow!" he says, "That's a very clever dog."
> "He's not that clever," replies one of the other players.
> "Every time he gets a good hand he wags his tail."

This clever dog couldn't hide his happiness and had to share it by wagging his tail. OK, it was a little bit of fictional fun but we all recognise that inherent quality in our dogs, how they share so much of themselves in such an easy and natural fashion.

In our own case, at the time[44] of writing this, we live with ten dogs, eight of whom are ex-rescue dogs. It would be reasonable to imagine that any dog that had come either straight off the street, a stray dog in other words, or from a dog-rescue centre, would have some behavioural issues. To a small extent this has been seen by us, that some dogs do come to us with a few minor, antisocial issues. Nevertheless, the way that our existing dogs in the home quickly assess and welcome a new dog to their family, how they instinctively know that they are going to fit in, is a model of openness and acceptance, topics that I write more on in the forthcoming chapters. Here, I want to stick specifically to sharing.

Sharing is synonymous with selflessness. We couldn't openly share a life in a selfless manner if it wasn't possible to push to the back of our mind, to the back of our consciousness, our need for self. In other words, diminish significantly our need to protect our ego. For the clear reason that if our egos dominate how we behave then selfless sharing would be very difficult, some might say more than very difficult but impossible.

A dog seems to know clearly and instinctively that its best interests depend on getting on with other dogs in a domestic environment, frequently those other dogs being of different breeds and from very different backgrounds. Inevitably, how a dog manages those boundaries of sharing, from the perception of the dog, is beyond the understanding of dog owners. So we observe how dogs will lick each other, snuggle up and sleep together, play together and share, in what clearly appears to be a trustful, loving community, without understanding the psychological drivers within those same dogs.

Nonetheless, that natural sharing that we see in our dogs links effortlessly with our human need for sharing. I had to look up and remind myself who it was that coined the expression: "No man is an island." It was the English poet John Dunne[45]. A beautiful reflection on our human need for sharing.

There are numerous benefits for having a dog or two in our life but possibly the core benefit is the one of never feeling alone. Think how often one sees a homeless person by the side of the road begging for food, money or for a lift, and nearby is their dog. Irrespective of the certainty that being homeless is tough, is the added certainty that it is a great deal tougher if there is also a dog to feed and look after as well. My strong sense is that the sharing of the lives of two creatures, human and dog, more than offsets the added challenges of having a dog in your life if you have no permanent home.

No better underlined than by an article seen on the website of *Flagpole Magazine*[46], the "locally owned, independent voice of Athens, Northeast Georgia."

The article[47] was titled: "Dogs and Their Homeless Owners Share Love, If Not Shelter," and was written by Stephanie Talmadge. It opened:

> If you walk down Clayton Street, specifically near the College Avenue intersection, you may have received a furry greeting from a little brown, scraggly pup. Usually a blur, due to near-constant wagging, this tiny dog, Malika, spends many of her days guarding that corner for her owners, David and Dorothy Gardener, who are experiencing temporary homelessness.
>
> Though the Gardeners are homeless, little Malika is far from it. She's not in the pound, waiting to be adopted or rescued before her time runs out. She's not running around in the streets or woods, fending for herself.

Later on Stephanie Talmadge confirms my earlier view that a homeless person benefits from having a dog. This is what she writes:

> Homeless or not, owning a pet is a huge responsibility, and obviously it can be extremely rewarding, well worth the complications it creates. Plus, a person doesn't have to be homeless to have financial barriers to providing good care. Plenty of dogs who live in permanent housing are neglected and mistreated daily.
>
> "Just because someone's homeless shouldn't mean they're not allowed to have a companion animal," Athens-Clarke County Animal Control Superintendent Patrick Rives says, "And there may be some good reasons for them to [have one] . . . There is a psychological impact of having a companion animal, and I wouldn't want to take that away from someone."

Going back to an earlier time, about 1870, Senator George C. Vest delivered a powerful and moving eulogy about the dog, addressed to the jury at the Old Johnson County Courthouse in Warrensburg, Missouri. It was in response to his client's dog, Old Drum, having being shot and killed the previous year. Here is that eulogy. You will get more benefit from the words if you can read them slowly aloud:

> The best friend a man has in the world may turn against him and become his enemy. His son, or daughter, that he has reared with loving care, may prove ungrateful. Those who are nearest and dearest to us, those

whom we trust with our happiness and good name may become traitors to their faith. The money a man has he may lose. It flies away from him, perhaps when he needs it most. A man's reputation may be sacrificed in a moment of ill-considered action. The people who are prone to fall on their knees when success is with us may be the first to throw the stone of malice when failure settles its cloud upon our heads.

The one absolutely unselfish friend that man can have in this selfish world, the one that never deserts him, the one that never proves ungrateful or treacherous, is his dog. A man's dog stands by him in prosperity and poverty, in health and in sickness. He will sleep on the cold ground when the wintry winds blow and the snow drives fiercely, if only to be near his master's side. He will kiss the hand that has no food to offer, he will lick the wounds and sores that come in encounters with the roughness of the world. He guards the sleep of his pauper master as if he were a prince. When all other friends desert, he remains. When riches take wing, and reputation falls to pieces, he is as constant in his love as the sun in its journey through the heavens.

If fortune drives his master forth, an outcast in the world, friendless and homeless, the faithful dog asks no higher privilege than that of accompanying him, to guard him

> against danger, to fight against his enemies. And when that last scene of all comes, and death takes his master in its embrace and his body is laid away in the cold ground, no matter if all other friends pursue their way, there, by the graveside will the noble dog be found, his head between his paws, his eyes sad, but open in alert watchfulness, faithful and true, even in death.

"The one absolutely unselfish friend that man can have in this selfish world is his dog." Words that are as true today as they were in 1870. By the way, Vest won the case for his client!

Continuing on from the sharing that we see between dogs, and between dogs and humans, what do we know and understand about how we share amongst ourselves? There is no shortage of examples of humans engaged in wonderful acts of sharing amongst themselves. I shall dip into just three of those examples.

Eric Berger, the science writer for the *Houston Chronicle*, wrote an article[48] in March 2012 under the heading of: "What Makes Us Human? Teaching, Learning and Sharing."

He opens the article by reminding his readers that in the briefest of time, compared to the long history of planet Earth, humans, "have gone from subsistence as scattered, hunted bands of cave dwellers to dominating the planet."

Later, Berger quotes Steven Schapiro, a professor at the University of Texas MD Anderson Cancer Center's primate research facility in Bastrop, and home to 169 chimpanzees, saying, "We wanted to understand why our closest living relatives can't do all of the kinds of things we do."

Berger then explains: "To address their question the

scientists devised a series of puzzles with escalating difficulty, the solving of which would produce rewards – stickers of increasing attractiveness for kids; carrots, apples and then grapes for the monkeys."

Those scientists reported the following:

> During the experiment the researchers observed that the children treated the puzzles as a social exercise, working them together and giving verbal instruction to one another. When successful, they shared the rewards.
>
> In contrast the chimpanzees and capuchins appeared to only see the puzzles as a means to obtain rewards, and worked mostly independently and did not learn from their efforts. They never shared.
>
> Humans, then, have ratcheted up their culture by teaching one another, imitating the successful behaviours of others and altruism.

"When successful, they shared the rewards."

Who knows if perhaps when we humans uniquely had interactions with evolved wolves, later to be known as dogs, in our lives thousands of years ago (back in the times when our survival depended on effective hunting and gathering) if learning to share that hunting and gathering with one other and our dogs, embedded within us the virtue of the sharing of rewards? I would like to think so.

My last two examples, ones that will end this chapter, are two wonderful modern examples of a sharing culture, regrettably not referring to dogs. The first of these two is the Buy Nothing Project[49] that has as its subheading: Random Acts of Kindness All Day Long.

As the website[50] explains:

> **Buy Nothing. Give Freely. Share Creatively.** The Buy Nothing Project began as an experimental hyper-local gift economy on Bainbridge Island, WA; in just 18 months, it has become a social movement, growing to over 110,000 members in 12 nations with 607 groups and 800 volunteers. Our local groups form gift economies that are complementary and parallel to local cash economies; whether people join because they'd like to quickly get rid of things that are cluttering their lives, or simply to save money by getting things for free, they quickly discover that our groups are not just another free recycling platform. A gift economy's real wealth is the people involved and the web of connections that forms to support them. Time and again, members of our groups find themselves spending more and more time interacting in our groups, finding new ways to give back to the community that has brought humor, entertainment, and yes, free stuff into their lives. The Buy Nothing Project is about setting the scarcity model of our cash economy aside in favor of creatively and collaboratively sharing the abundance around us.
>
> ***"It has become a social movement collaboratively sharing the abundance around us."***

My final example is about as different to the previous one as you could imagine. It is an example taken from the software industry. More specifically from what is called the Open Source Initiative. As the home page[51] of their website explains: [My bold emphasis] "Open source software is software that can be freely used, changed, and ***shared*** (in modified or unmodified form) by anyone."

Now for many people, the phrase "open source software" might bring on a couple of yawns but the reality of today's interconnected world is a direct result of such open source software.

There was an article written by George Bradt in the November issue[52] of *Forbes Magazine*. It was titled "Why Open Leadership Has Become Essential." Just read the opening paragraph.

> You would not be reading this if open source software did not exist. Without open source standards, the Internet would not exist. This article would not exist. Those of you whose parents met on Match.com would not exist. All of you should be thankful for open source software. Now, as the world has changed, open source software's principles of openness, transparency and meritocracy have become essential standards for leadership in general.

So what better way to end this chapter on sharing than by highlighting a section of that last sentence:

"Principles of openness, transparency and meritocracy have become essential standards for leadership in general."

Sharing seems like the way to go forward, not just for our leaders but for all of humankind.

If I was a dog, it would be impossible to stop my tail from wagging.

Chapter 19

HONESTY

What I am about to say probably applies to all animals but I have in mind one specific animal that won't surprise you: the dog. Namely, that the dog is incapable of guile or deceit. Thus a chapter about the quality of honesty in dogs might seem a little bizarre. Because honesty, and dishonesty, surely are terms that exclusively describe human tendencies.

When the term "dishonest" is used to describe a person, most often we are describing an effort by that person to deceive another. It is a description of someone who is intentionally trying to mislead or misinform another person.

Now that last statement is not to imply that dogs don't try to manipulate us humans, far from it. Their attempts at manipulation would impress any three-year-old child. But there's nothing dishonest about a dog trying to manipulate its owner into giving the dog whatever it wants for they are far too obvious in their motives and methods. As I once read in "A Dog, Honestly" by Eric Brad, dogs are: "Just scavengers looking for a way to get something with minimal effort."[53] Thus I think we can take it as a given that dogs are fundamentally honest creatures.

Dear reader, you have no way of knowing that after writing that last sentence, I sat staring at the screen for a good ten minutes. I didn't know how to continue the chapter. I couldn't think of anything to add to what every person knows, and has known since time immemorial, honesty is a fundamental aspect of being a good person, "the most enviable of all titles", as George Washington wrote in a 1788 letter.[54]

What was exercising my brain was to come at the subject of honesty in a way that offered a compelling reason for being honest, over and above the natural assumption that honesty is good and so blindingly obvious not to require being spelled out. For I had the sense that honesty, sitting on the bedrock of integrity, is very different to the other qualities that we need to learn from our dogs. Because honesty is a fundamental way of relating to the world around us (ignoring the question of "white lies")! Whereas such attributes that I write about in the chapters to follow, qualities such as forgiveness, openness, acceptance, and adaptability, are perfectly capable of being learned.

But, at heart, I was still lost as to how to proceed with the chapter.

So, I spent another thirty minutes exploring the Web looking for clarity, or more honestly written as me looking for some inspiration. Yet those Web searches just ended up with me becoming more confused. About the least confusing item I came across, more or less at random, was an article on a website *The New Atlantis*. The full article, written by Lee Perlman, was titled: "The Truth about Human Nature".[55] Here's an extract:

> Since Nietzsche, the choice of which version
> of ourselves we identify with has been widely

> understood as a choice between lying and truth-telling, to ourselves as much as to others. The moral ideal has become authenticity, a particular kind of honesty. Of course, just about any philosophical ideal is grounded in some sort of honesty: the search for Truth requires truth. Yet Aristotle describes honesty as a virtue only of self-presentation, the balance between self-deprecation and boastfulness. And Plato never lists honesty as a virtue at all, and even distinguishes between "true lies" and useful or noble lies. From the modern to the post-modern era, honesty and authenticity shifted to become much of the *telos*[56] of life, where before they had been but means in our progress toward that end.

But that extract seemed to confirm in my mind that honesty, something that by rights should be fundamentally understood, was anything but simple. What to make of all this?

I am going to fall back on the ideas expressed in chapter 16 about community. Specifically on the closing words of that chapter, "For one thing is clear beyond doubt. Humans are not designed to be alone; isolation and loneliness are damaging to our mental health. One might go so far as to say that the rewards of sharing, of living a local community life, might just possibly be the difference between the failure and survival of the human race. Now that is something we need to learn from our dogs."

Ex-rescue dog Casey sitting on Jean's lap, after his rescue from an animal shelter he had been in for a year, where he was due for euthanisation.

Chapter 20

FORGIVENESS

Dogs offer many examples of forgiveness that many won't see clearly in human terms.

Think of those dogs that have been treated cruelly, often for months or years, and then find themselves in a new loving home. Think of dogs that have spent weeks and months in confinement at the local humane centre. Or more terrible to comprehend are those dogs that have simply been abandoned, just thrown away by a so-called human.

We take it totally for granted that when dogs find that new loving home that they will adjust quickly and easily. Let me write about Casey, one of the dogs that we have at home. He was found in the local dog rescue unit in Payson, Arizona, when we were living in that part of America, before we moved to southern Oregon. Casey had been confined in that dog rescue unit for a year and almost certainly hadn't found a new home because he was a Pit Bull mix, and looked it! Two weeks before he was due to be put down, now classified by the rescue unit as being impossible to adopt, Jean brought him home.

The speed at which he settled in to his new home including, not that long after, a house move from Payson,

Arizona, to Merlin, Oregon, was just wonderful. Never for a moment did Casey display any cautiousness or nervousness towards Jean and me, and, even more important, didn't reveal any antisocial inclinations towards visitors who came to the house. Casey has a wonderful temperament and is a happy, lively, and affectionate dog. Clearly, Casey harbours no grudges from his past experiences. His forgiveness of the way his life previously had been dominated by the actions of humans is flawless. With Payson in mind, there's another example of how a dog so quickly puts past experiences behind them and embraces their new life. For in Payson we knew the author, Trish Iles, who has the blog "Contemplating Happiness".[57] Here is a lovely story from Trish, in her own words:

> **What the Dog Knew**
> I was pondering the eternal question: why does two weeks of relaxing vacation seem like so much more time than two weeks of working like my pants are on fire, here at my desk? My sweet husband and I talked about it a little bit, but came to no definitive answer. I chatted with friends about it. No insights. Google had no opinion, either.
>
> Chloe came to us from a rescue organization. I think sometimes about what her experiences have been in her young life. She started out as an abandoned puppy on a reservation in New Mexico and was soon in the pound where she was on the euthanasia list. A kind woman rescued her and took care of her until she found us: just when Chloe was becoming at home with the

> rescue lady, she was uprooted again and sent home with two new people. What must she have been thinking?
>
> Chloe didn't close her heart to us, though. She watched for a few days. When she decided we weren't going to make dinner out of her and that she was really staying with us, she threw her whole being into becoming one of the family. She let herself trust us.
>
> I'm not sure I would have had the courage to trust a new set of people again. I'm doubly not sure that I give a rat's patootie what those new people thought of or wanted from me. Chloe was willing not only to trust us, but to love us. She forgave us immediately for ripping her from the home she knew, and she adopted us right back.
>
> Chloe was born knowing. She knows about joy. She knows about living a life in balance. She knows about forgiveness, trust, exuberance, a passion for learning and the power of a good nap. I think that when I grow up, I want to be just like her.

When I think of the many dogs that we have in our home that have endured terrible conditions (starvation, abuse and worse) before they were rescued, it is beyond my understanding that they have become such loving creatures so quickly after they were in a safe environment

Much too late to make me realise the inadequacies of my own parenting skills, I learned an important lesson when training Pharaoh (the first dog I had ever had) as a puppy back in 2003. What I learned was that putting more emphasis into praise and reward for getting it right "trains" the dog much more quickly than telling it off. The classic example is scolding a dog for running off when instead there should be lots of hugs and praise for the dog returning home. The scolding simply teaches the dog that returning home isn't pleasant whereas praise reinforces the dog's belief that home is the place to be. Summed up by the well-known phrase: "Catch them in the act of doing right."

Like so many things in life, that concept of praise over punishment is so very obvious once understood. This approach, this philosophy, works with youngsters in just the same positive way.

Let's look at the nature of forgiveness in humans. But before doing so, I must make something clear to you, and that is the balance of this chapter predominantly comes from detailed research undertaken by me with two psychotherapists and a professional counsellor. What I learned from these good people was also supported by research elsewhere. Nevertheless, if any reader feels affected by this chapter then they should seek the appropriate assistance, preferably by first contacting a national or regional organisation that can identify a professional person convenient and appropriate to your needs.

There are essentially two options that we can choose to act out when we are hurt by someone. We can hang on to those feelings of anger and resentment, possibly having thoughts of revenge, or we can respond with forgiveness. The first leads to wounds of anger, bitterness and resentment. The second leads to healing, to the rewards of peace, hope, gratitude, and joy.

The powerfully positive outcome from acting with

forgiveness is that the act that caused the hurt is denied the chance of producing any real emotional force. You quickly put it behind you and focus on other, more positive parts of your life. That's not to say that the hurtful act is forgotten, possibly not for some time if it was a significant hurt, it is just that it greatly lessens its grip on you. Frequently, forgiveness can stimulate feelings of understanding, empathy and compassion for the person who hurt you.

Forgiveness also doesn't mean that you deny the other person's responsibility for causing you to be hurt nor does it minimise, let alone justify, the wrongness of the act. The key is that the person can be forgiven, *without excusing the act*. Think of it as forgiveness bringing a kind of peace that helps you get on with your life.

Actually, the benefits of forgiveness are even more tangible than the subjective nature of the bringing of peace. For there is clear evidence that shows that letting go of grudges and bitterness, of offering forgiveness in other words, can enable a string of benefits:

- Healthier relationships
- Greater spiritual and psychological well-being
- Less anxiety, stress and hostility
- Lower blood pressure
- Fewer symptoms of depression
- Stronger immune system
- Improved heart health
- Higher self-esteem

And if that doesn't beat a few bottles of pills, then I don't know what does.

I paused at this point and read back to myself those last few sentences. They struck me as having the slight tone of a sermon. That what needs to be added to those ideas is how we learn to forgive in a practical manner.

Psychotherapists and others from similar backgrounds say that forgiveness is the result of change or more accurately put, a commitment to a process of change.

To put some flesh on the bones of that last point, that forgiveness results from a commitment to a process of change, what now follows are five recommendations typically offered by psychotherapists and professional counsellors. They would recommend resisting the temptation to read through all of them without pause. Very simply, I am told, because the value of each recommendation is enhanced if you allow sufficient time for your mind fully to embrace each meaning. Or, as one psychotherapist put it to me, giving your heart time to engage with your head.

I chose, therefore, to separate each recommendation with a small literary device to prompt such a pause.

But first, before moving on to the five recommendations, we need to refocus on what hurt us. So go and find a place where you can sit quietly and give your mind the space to bring up an episode where someone caused you hurt. It could be a recent episode or one from long ago that still has the potential to hurt you. Dwell on it without mental interruptions.

Recommendation number one: Consider the values of forgiveness to you, not at a theoretical level, but to you in terms of where you are at this point in your life and, by implication, how important those values of forgiveness are to you at this given time. When you are ready, read on.

Look away from the page. Reflect on what that means. Think it with your head and feel it with your heart. Only when you feel you are ready, move on to the second recommendation.

Recommendation number two: Reflect on the real facts of what happened, how it came about, how you reacted, and to what degree the situation has affected, or has the potential to affect you, your life, health, and your well-being.

Look away from the page. Reflect on what that means. Think it with your head and feel it with your heart. Only when you feel you are ready, move on to the third recommendation.

Recommendation number three: Actively choose to forgive the person who hurt you. Possibly by re-reading this chapter down to this point. Actively choose to forgive that person right now. Stay with that forgiveness.

Now look away from the page. Reflect on what that means. Think it with your head and feel it with your heart. When you feel you are ready, move on to the fourth recommendation.

Recommendation number four: the penultimate recommendation: Stop seeing yourself as a victim of the hurtful event. Understand that by continuing to feel victimised, you will be unable to release the control, the power that the offending person and the situation has over you at this present time.

Look away from the page once again. Reflect on what the recommendation means. Think it with your head and feel it with your heart. Only when you feel you are ready, move on to the final recommendation.

Recommendation number five: As you let go of the pain, let go of the hurt and your grudges. Your life is now no longer defined by how you have been hurt. Better than that, much better, is that the letting go opens your heart to finding compassion and understanding for the other person.

Once more, look away from the page. Reflect on what that letting go means. Think it with your head and feel it with your heart. When you feel perfectly ready, read on.

There is no question that forgiveness can be challenging at times, especially if the person who's hurt you doesn't admit he or she is wrong or doesn't ever speak of his or her sorrow. However, never allow yourself to become stuck. Reflect on the times when you have hurt others and when those others have forgiven you. Share your burden of finding forgiveness, such as writing in a journal, or through prayer or guided meditation. Even better, open up to someone you've found to be wise and compassionate, such as a spiritual leader, a mental health provider, or an impartial loved one or friend. And share your burden with your dog. They are such great listeners.

Never forget that finding forgiveness is a process and even small hurts might need to be forgiven over and over again.

In the vast majority of cases, forgiveness can lead to reconciliation. Especially so when the hurtful event involved someone whose relationship you really value, for example, someone emotionally close to you. However, there is one case where reconciliation is impossible. That is the case where the person who hurt you has died. Even then, when reconciliation isn't possible, forgiveness still is.

Also don't forget to respect yourself, to keep an open heart and mind and do what seems best for you in the specific situation.

The final thought for this chapter on forgiveness is this. Do not think that it is about the other person needing to change, that isn't the point of forgiveness. Forgiveness is about us, how it can change *our* lives through bringing peace, happiness, and emotional and spiritual healing. It also helps, enormously so, in allowing us to recognise our own faults, our own mistakes, and the times when we have hurt others, so that we can offer our apologies in an open and honest manner.

Dogs so easily forgive the many ways in which we humans accidentally, or otherwise, hurt them. Forgiveness is one of the most precious of all the many precious qualities that we can learn from our dogs.

But there's a postscript. There is an appendix in this book: A Horse Called Ben. Do read it because it demonstrates both the appalling cruelty that some so-called humans can dole out to animals, in this case a horse. It also reveals the quality of forgiveness that Ben demonstrated so quickly.

Chapter 21

OPENNESS

Like so many others, I am guilty of using terms and words, and never reflecting on exactly what that term or word truly means. (That discovery was an unanticipated personal benefit of writing this book.) So it is with the title of this chapter: openness. I searched around for some meaning, some idea of what openness really meant. It's such a little word to run off the tongue, seemingly so easy to understand the meaning of the word, that it isn't until one pauses and asks oneself, "Do I really, truly know what the word openness means, what it really conveys?" that the doubt creeps in.

The dictionary didn't help that much either. "**Openness** *noun* **1.** The condition of being laid open to something undesirable or injurious. **2.** Ready acceptance of often new suggestions, ideas, influences, or opinions."[58] No, that still didn't offer to my way of thinking a clear message about the meaning of the word.

Then my eye wandered up the dictionary page to the entry immediately above openness; to "**open-mindedness**: *noun* Ready acceptance of often new suggestions, ideas, influences, or opinions : openness, receptiveness, receptivity, responsiveness."[59] Bingo: responsiveness and receptiveness! Those qualities did speak to me because they were the qualities of openness that I see in dogs.

Dogs don't appear to engage in introspection. They don't seem to worry about who they are. Their emotions are clear to us. One might say that dogs wear their emotions on their paws. They engage with the human world around them more-or-less regardless of our moods, our personal situations, and choices. We call them and expect them to come. Perhaps, ask them to go to a part of the house if we are going out, or to stay in a place until we tell them they can move. Dogs appear simply to be there for us, being there for us only on our terms. As much as each day is unique and different, dogs offer constancy, a reliability, that feels unmatched in us.

Jean and I depend on our dogs, as do the vast majority of people who have dogs in their lives. They calm us in times of trouble, give us a better perspective of life's "big picture". We can so openly share a sense of joy with our dogs. Dogs give us permission to be silly with them, to hug them, to rub their tummies, or to roll around on the floor with them. It is possible, easily so, to learn something from a dog every single day simply from observing closely how these beautiful animals live. In return, dogs ask only of us for food, for water, and for affection.

The openness of dogs has been celebrated in many ways for centuries, both in song and verse. It is still to be celebrated today, for today that wonderful quality of openness is still alive and vibrant in our dogs. It seems so much more than just the product of some evolution of nature. Reflect on the incredible range of dogs, on all that selective breeding, on the many differences in the environments in which dogs live out their lives. So many dogs and yet every one of them coming to us, to meet us, to be with us, just as they are, with no apologies and no covert agendas. As author Susan Kennedy[60] once said, "Dogs are miracles with paws."

I have had a dog in my life, my beloved Pharaoh, since 2003. I have had a great number of dogs in my life since meeting Jean in 2007, as many as fourteen and ten at the time of writing[61] these words. I can't imagine my life without our dogs. They truly provide unconditional love and they do so without hesitation. It is a simple, yet immensely beautiful relationship. The love that we receive from our dogs comes from their openness. A dog's openness is their gift. A precious, remarkable gift.

Now how on earth can one translate that as a quality that we humans have to learn from our dogs? Are there any practical benefits for us in practicing the openness we see in dogs? By raising the question, I am, of course, admitting some degree of doubt about being able to answer that question in the affirmative. Not doubting that there are benefits, just unclear about how to describe them. Unclear how we could ever match the openness of dogs.

So what I am going to do is to flip the issue on its head. Just stay with me a little longer.

I have referred to J previously in this book. In his world of counselling and therapy, J speaks like this. Namely, that in the world of solutions-focussed therapy, the area that J practices in professionally, the way forward with the person who has come to see J is always to focus "on what is working". He explains that while one would initially allow the problems to be voiced, this negativity would always be a tiny piece of the overall process, say less than 5 percent of the session. That even if a client's whole world seemed to be failing, there would always be something that was right, always a 1 percent that was working, and that would be the place to start.

No better endorsed than by the website of the organisation "Good Therapy".[62] I quote [my bold emphasis]:

> Solution focused brief therapy (SFBT[63]) targets the **desired outcome of therapy as a solution** rather than focusing on the symptoms or issues that brought someone to therapy. This technique only gives attention to the present and the future desires of the client, rather than focusing on the past experiences. The therapist encourages the client **to imagine their future as they want it to be** and then the therapist and client collaborate on a series of steps to achieve that goal.

Returning to the example of openness that we see in our dogs then maybe rather than wringing our hands because we will never be as open as those wonderful animals, perhaps we should flip the idea on its head. Ergo, not striving to be the same as our dogs, just trying to follow their example.

That is, for every one of us to be more mindful of the need for openness, to practice openness as a conscious idea, and to develop the habits of openness. No better summarised than those words of meaning from the dictionary a few paragraphs back:

"Ready acceptance of often new suggestions, ideas, influences, or opinions."

Seeing our dogs as a marvellous pillar, a wonderful example, of the greater openness that we all seek.

Chapter 22

ACCEPTANCE

When I was deliberating the list of dogs' qualities that we humans need to emulate, I was unsure if this chapter on acceptance, together with the previous one on openness, and the next one on adaptability, weren't all too close to the same idea. But after having thought about it some more, I decided that these three qualities are sufficiently different to warrant them being three specific chapters.

For openness is all about offering ourselves to others, whereas acceptance is, in a manner of speaking, the reverse current, allowing the outside world to flow in to us without too many mental and emotional "filters" corrupting that inward flow. And adaptability is all about change.

What example do our dogs offer us when it comes to acceptance? Almost immediately comes the answer: dogs accept the humans around them and the human world, accept their life as a domesticated animal, as a pet in other words, accept it all as it is for what it is, and calmly so.

Just think for a moment of the vast range of life experiences that our dogs live with. From the tiniest poodle who rarely is separated from its owner, to the sheepdog that works the land and spends its nights outside in the barn, all the way to the German Shepherd guard dog that is hardly a pet. Dogs are authentic, in the fullest meaning of the word. They react to their environment and to their natural instincts but totally within the human world in which each particular dog has been cast.

That is a level of acceptance that we humans can only dream about.

Nevertheless, even if that level of acceptance of the world about us by our dogs is most likely beyond human reach, there is still an important lesson to be had.

Let me elaborate.

The quality of the relationships that we have with other people revolves entirely around how we view them. And nowhere is that more important than how we view our loved ones namely, our family, our spouses, our partners, and our friends.

If we use the wonderful way in which dogs accept their world and, most notably, the way they accept other dogs (in general, I should add), as a model for the way we should accept our loved ones, there is much research to underpin the fact that, in consequence, we will enjoy wonderful relationships. If we quietly admit to ourselves that we do not accept our spouses and partners as fully as we should, then learning to accept them fully will miraculously transform our relationships. (Yes, it's a subjective generalisation from this fellow traveller of life, yet I hold to it being a valid statement.)

For the acceptance of people you share your life with is the greatest gift you can give them. It underlines how much you love them and how much you respect them. It demonstrates that you know that the decisions they make, from small ones to large ones, are based on what they believe is right. It doesn't at all prevent you from offering support and guidance, of course not. What it does guarantee is that you don't stray into criticism, especially the genre of criticism that has its roots in your (false) belief that other people are not thinking like you, not seeing something as you see it. For one very obvious reason: they aren't you.

There is no question at all that acceptance is the greatest gift you can offer someone, especially someone emotionally close to you, because it is the greatest sign of respect. And respect is the cousin of trust and without trust there is no relationship. It applies equally to humans and dogs. Just because we accept our dog unconditionally, knowing that our dog is completely authentic, because we know that it is a dog and never expect it to be anything other than a dog, doesn't in any way mean that the same approach, the same unconditional acceptance of a person in our lives, should not be our way of living with that other person.

Chapter 23

ADAPTABILITY

Apparently, about 200 years ago, around 1790 to 1800, the word "able" was added to the word adapt to make the word adaptable. From that came the related word, adaptability, the title of this chapter.

Our history, the long history of humankind, reveals a species that is incredibly adaptable. Yet, my sense of how adaptable any one person might be is inextricably wound up with change, and change is often a bitter fruit to taste.

You might recall that I opened chapter 11 "On Change", in Part Three of this book, with a spoken line from the film *Interstellar*: "They didn't bring us here to change the past." There's another spoken line from the film that I recall: "We all want to protect the world, but we don't want to change." That sentiment could be applied to so many aspects of our lives, especially to any form of change that heralded perceived uncertainty, or potential vulnerability; anything that might be regarded as taking us outside our comfort zone. Granted not every one of us.

Dogs, just like us humans, love routines. What strikes me from having lived for a number of years with, as you know, a great many dogs in the home is how amazingly a dog will easily adapt to new circumstances, both temporary and permanently changed circumstances.

Somewhere in my research, and I regret not being able to quote the reference, I came across a review of the author Jean Donaldson[64], in connection with her book *Culture Clash*. This book has shaped thinking about the behaviour of dogs and the relationship between dogs and humans.

The reviewer, in discussing the adaptability of dogs, proposed, "Maybe it's the simple way they view their world. Each thing in their lives seems to fall neatly into its place in their world view. Things to seek out, things to avoid, things to keep, and things to leave behind." Then a couple of sentences later, the reviewer added: "I would guess that scavengers need that kind of mind set. Take it as it comes, deal with it, and move on. Dogs seem to have developed a sense of adaptation. They see what needs to be done and simply find a way to do it no matter what the impediments might be." That last sentence describes a level of adaptability that, in my opinion, would be rare to find in a person.

My own experience of the adaptability of dogs is one way. By that I mean that my experience of dogs adapting to a new home is about dogs being rescued as homeless strays and being offered a loving home with Jean and me. The ease with which dogs adapt from being homeless to being with us is no better illustrated than by the story of Dhalia. Jean recounts how she found Dhalia.

> It was in 2005, about three months after Ben (my first husband) died. I was driving out to the small Mexican fishing port of La Manga

> where there were many stray dogs. The aim was to feed them on a regular basis and hope that they would become sufficiently comfortable with my presence so that they could be caught, so that they might be spayed or neutered and then offered for adoption.
>
> On the way there, I drove past a couple of dogs running alongside the highway. Dogs frequently did this looking for roadkill that they could feed on. I stopped the car wanting to put out some food and water.
>
> One of the dogs was so feral that it immediately took off into the bush. I turned around and the other dog was standing about ten feet away. It was cadaverous and obviously suffering from mange but cautiously came up to the food, sniffed carefully, and then started to eat. That dog allowed me to pick it up and then sat quietly with me on the front seat of the car while I continued to La Manga. It sensed immediately that it was safe and from that day has remained with me. I named her Dhalia.

When I joined Jean in Mexico and later when we ended up living in southern Oregon, Dhalia still showed her feral instincts when we went walking in the forest by constantly looking for food, despite the fact that she was by then a well-fed, happy and contented dog. Dhalia died in April 2014 and is still missed.

But there is a return journey that illustrates the wonderful adaptability of dogs. I have in mind the way that some dogs can so quickly adapt from living in a domesticated environment to living as homeless dogs.

The future is unknown, of course. But it is more likely to be delivering changes that are way beyond our expectations. Learning the degree of adaptability that our dogs demonstrate so very often will be critical to humankind. I play with this theme as a fictional short story in Chapter 26 of this book.

Dhalia riding on Jean's lap.

Chapter 24
STILLNESS

Stillness! It is a simple, single word yet, somehow, it sounds as though it belongs to a different age, as though stillness is a long way from the modern society that millions and millions of us subscribe to.

Stillness is the last quality that I want to write about, as in the last quality that I see in our dogs that we humans should emulate. There was a very deliberate reason to make it the last one. But, if you will forgive me, I'm not going to explain why until towards the end of this chapter.

The dog is the master of being still. Being still, either from just lying quietly watching the world go by, or being still from being fast asleep. The ease at which dogs can find a space on a settee, a carpeted corner of a room, the covers of a made-up bed, and stretch out and be still, simply beggars belief. Dogs offer us the most wonderful quality of stillness that we should all practice. Dogs reveal their wonderful relationship with stillness.

In August 2014, TED Talks published a presentation[65] by Pico Iyer. Pico Iyer, a British-born essayist and novelist of Indian origin and the author of a number of books about crossing cultures. He has been an essayist for *Time Magazine* since 1986. His TED Talk was called, "The Art of Stillness."

When I listened to Iyer's talk,[66] I found it utterly riveting. In a little over fifteen minutes, his talk touched on something that so many of us feel, probably even yearn for: the need for space and stillness in our minds. Stillness to offset the increasingly excessive busyness and distractions of our modern world. Or to use Iyer's words, "Almost everybody I know has this sense of overdosing on information and getting dizzy living at post-human speeds."[67]

Now Iyer has clearly been a great traveller and the list of countries and places he has visited is impressive. From Kyoto to Tibet, from Cuba to North Korea, from Bhutan to Easter Island, this is a man having grown up both being a part of, and yet apart from, the English, American, and Indian cultures. Yet of all the places Pico has been to, guess what he tops them all with? What he discovers in stillness: "that going *nowhere* was at least as exciting as going to Tibet or to Cuba" (my italics emphasis).[68]

Here are some of his words from that TED Talk. First he says:

> And by going nowhere, I mean nothing more intimidating than taking a few minutes out of every day or a few days out of every season, or even, as some people do, a few years out of a life in order to sit still long enough to find out what moves you most, to recall where your truest happiness lies and to remember that sometimes making a living and making a life point in opposite directions.

Then saying a few moments later:

> And of course, this is what wise beings through the centuries from every tradition have been telling us. It's an old idea. More than 2,000 years ago, the Stoics were reminding us it's not our experience that makes our lives, it's what we do with it. . . . And this has certainly been my experience as a traveler. Twenty-four years ago I took the most mind-bending trip across North Korea. But the trip lasted a few days. What I've done with it sitting still, going back to it in my head, trying to understand it, finding a place for it in my thinking, that's lasted 24 years already and will probably last a lifetime. The trip, in other words, gave me some amazing sights, but it's only sitting still that allows me to turn those into lasting insights. And I sometimes think that so much of our life takes place inside our heads, in memory or imagination or interpretation or speculation, that if I really want to change my life I might best begin by changing my mind.

Emails, smartphones, telephone handsets all around the house, television, junk mail on an almost daily basis, advertising in all its many forms, always lists of things to do, and on and on and on. There was a television documentary screened in the first half of 2015 that was a detailed examination about obesity, now reaching epidemic proportions in a number of countries. It included this statement: "Modern society has ended up being too busy to eat properly."

It's as if in this modern life, with so many wonderful ways of doing stuff, of connecting with others, being entertained, and more, we have forgotten how to do the one most basic and fundamental thing: Nothing. It's as if so many of us have lost sight of the greatest luxury of all: immersing ourselves in that empty space of doing nothing.

Thus it wouldn't be surprising to learn that people are becoming aware of this madness, that more and more are taking conscious and deliberate measures to open up a space inside their lives. Whether it is something as simple as listening to some music just before they go to sleep, because those who do notice that they sleep much better and wake up much refreshed, or taking technology "holidays" during the week, or attending yoga classes or enrolling on a course to learn transcendental meditation, there is a growing awareness that something in us is crying out for the sense of intimacy and depth that we get from people who take the time and trouble to sit still, to go nowhere.

Here's a true story that is a poignant example of the stillness of a dog. Jean's first husband, Ben, became terminally ill in the early months of 2005. Towards the end of his days, when Jean could no longer share Ben's bed, one of their ex-rescue dogs, Lilly, took to being very still next to Ben for every hour of the day, only leaving his side to pee and dump and to hastily eat some food. Jean recalls that Lilly started being still beside Ben about three weeks before he died. That was when Ben lost any awareness that Jean was in the room with him but, as Jean recalled, Ben was still conscious of Lilly being next to him as he would lean his hand across the dog.

Ben died in the early hours of July 10, 2005. When Jean awoke in the morning she immediately knew, before even getting out of her bed, that Ben had died during that night. For Lilly was perfectly still, quietly sleeping next to Jean.

Science supports the benefits of stillness, of slowing down the brain. In an article[69] posted on the *Big Think* blog, author Steven Kotler explains what are called "flow states": "a person in flow obtains the ability to keenly hone their focus on the task at hand so that everything else disappears."

Elaborating in the next paragraph, Kotler continues:

> So our sense of self, our sense of self-consciousness, they vanish. Time dilates which means sometimes it slows down. You get that freeze frame effect familiar to any of you who have seen The Matrix or been in a car crash. Sometimes it speeds up and five hours will pass by in like five minutes. And throughout, all aspects of performance, mental and physical, go through the roof.

The part of our brain, known as the prefrontal cortex, houses our higher cognitive functions, such as our sense of morality, our sense of will, and our sense of self. It is also that part of our brain that calculates time. When we experience flow states or what Steven Kotler describes as "transient hypofrontality" we lose track of time, lose our grip on assessing the past, present, and future. In Kotler's words: "We're plunged into what researchers call the deep now."

Kotler then explains:

> So what causes transient hypofrontality? It was once assumed that flow states are an affliction reserved only for schizophrenics and drug addicts, but in the early 2000s a researcher named Aaron Dietrich realized

> that transient hypofrontality underpins every altered state, from dreaming to mindfulness to psychedelic trips and everything in between. Sometimes these altered states involve other parts of the brain shutting down. For example, when the dorsolateral prefrontal cortex disconnects, your sense of self-doubt and the brain's inner critic get silenced. This results in boosted states of confidence and creativity.

Don't worry about the technical terms, just go back and re-read that last sentence: "This results in boosted states of confidence and creativity." Such is the power of stillness.

Back to Iyer and his concluding words:

> So, in an age of acceleration, nothing can be more exhilarating than going slow. And in an age of distraction, nothing is so luxurious as paying attention. And in an age of constant movement, nothing is so urgent as sitting still. So you can go on your next vacation to Paris or Hawaii, or New Orleans; I bet you'll have a wonderful time. But, if you want to come back home alive and full of fresh hope, in love with the world, I think you might want to try considering going nowhere.

At the start of this chapter, I wrote that I would leave it until the end to explain why I deliberately made this chapter on stillness the last one in the series of dog qualities we humans have to emulate, to learn as it were.

For the fundamental reason that it is the stillness of mind (that we so beautifully experience when we hug our dog, or close our eyes and bury our face in our dog's warm fur) that, like any profound spiritual experience, has the power to transform our mind from negative to positive, from disturbed to peaceful, and from unhappy to happy. The power of overcoming negative minds and cultivating constructive thoughts, of experiencing transforming meditations is right next to us in the souls of our dogs.

"So the darkness shall be the light,
and the stillness the dancing."
T.S.Eliot

Jean with Lilly, 2014 (Lilly died in 2015).

Ruby sitting behind Casey.

Part Five:

Reflections

Lilly, 2014.

Jean loving upon Paloma with Lilly just visible behind Paloma.

Chapter 25

A WAY INTO OUR OWN SOUL

"Happiness resides not in possessions, and not in gold, happiness dwells in the soul."

So wrote the philosopher Democritus who was born in 460 BCE (although some claim his year of birth was 490 BCE). He acquired fame with his knowledge of the natural phenomena that existed in those times and history writes that he preferred a contemplative life to an active life, spending much of his life in solitude. The fact that he lived to beyond 100 suggests his philosophy didn't do him any harm.

Now the last thing I am going to attempt is any rational, or even semi-rational, explanation of the soul: of what it is, of whatever it is. Despite the familiarity of the word, especially within religious circles, the notion of the soul remains an enigma. Indeed, it reminds me of that clever quotation attributed to the German philosopher, Martin Heidegger: "Making itself intelligible is suicide for philosophy". With a little poetic licence that might be rewritten: "In making itself intelligible, does the soul become soulless?"

Thus having failed the test of knowing rationally what a soul is I shall, nonetheless, continue to use the word. Because there will be sufficient connection between me writing the word "soul" and those reading the word for you, dear readers, to sense where I am coming from.

I'm going to stay with this wonderful concept of soul before adding our beautiful dogs into the dream. Or rather I am going to stay with the concept of soul courtesy of the writer John O'Donohue. John's name is not one known to the masses. Yet his writings are, without fail, beautifully moving. John's first book was titled *Anam Cara*, which means "soul friend" in the old Celtic language. The following passage, taken from *Anam Cara: A Book of Celtic Wisdom,*[70] represents to my mind the most exquisite understanding of the human soul.

> The secret heart of time is change and growth. Each new experience that awakens in you adds to your soul and deepens your memory. The person is always a nomad, journeying from threshold to threshold, into ever different experiences. In each new experience, another dimension of the soul unfolds. It is no wonder that from ancient times the human person has been understood as a wanderer. Traditionally, these wanderers traversed foreign territories and unknown places. Yet, Stanislavsky, the Russian dramatist and thinker, said that: "the longest and most exciting journey is the journey inwards."
>
> There is a beautiful complexity of growth within the human soul. In order to glimpse this, it is helpful to visualise the mind as a tower of windows. Sadly, many people

> remain trapped at the one window, looking out every day at the same scene in the same way. Real growth is experienced when you draw back from that one window, turn, and walk around the inner tower of the soul and see all the different windows that await your gaze. Through these different windows, you can see new vistas of possibility, presence, and creativity. Complacency, habit, and blindness often prevent you from feeling your life. So much depends on the frame of vision, the window through which you look.

Aren't those wonderful words from O'Donohue and a brilliant example of his exquisite creativity of thought. They also offer the most perfect window into seeing how the dog offers us a way into our own human soul.

What do I mean by this?

When we have dogs in our lives, there are many occasions when there is a link between us and our dog, a link that defies logical explanation.

For instance, as a human, out of the blue, with no rhyme or reason, you will surely experience finding your day a bit tough from time to time. The odds are that it doesn't show to your loved ones and you will be fairly certain that your experience of finding things a bit rough is well hidden inside you.

But you and I know you can't hide it from your dog. You slump down in a chair and your dog comes over and lays its warm snout across your legs, or offers a head for you to scratch. In any one of many familiar ways you have a caressing and loving contact with your dog. And you know, you know beyond doubt, that your dog is attracting the angst away from you.

Or how about the time when you might be standing somewhere in or around the house, trying to think how best to approach a task, and your dog comes along and softly leans against you. Or what about that most special of links between us and our dog. I have in mind the times when our dog links eye-to-eye with us, when those beautiful, deep unblinking eyes of our dog look so deeply inside us. Those are the times when you and your dog know, in a clear unwritten language, the thousands of years of a special relationship that humans and dogs have had, and still have, with each other. That at that moment of held eye contact there is a real, tangible connection between your two souls, between your human soul and the soul of your dog.

We know beyond doubt that dogs have emotions, that they are full of natural goodness and feelings, and that there is some part of a dog's inner being that links to us and, in turn, that there is an inner being within us that links back to our dog. Let me return to the power of that eye-to-eye bond.

In humans, that part of the brain in which self-awareness is thought to arise is called the ventromedial prefrontal cortex. Apparently, that just happens to be located behind the eyes. Ergo, we learn[71] to associate the identity of others with our eyes. Then as we mature, our eyes take on more importance because we develop awareness and a better understanding of the social cues that other people convey with their eyes.

Therefore, is it any surprise that dogs, being the intuitive creatures that they are, soon learn to read us humans and the feelings and emotions that we transmit from our eyes. There's a knowing in my mind, albeit an unscientific knowing, that dogs, too, give out emotions and feelings from their own eyes.

That loving a dog and being loved back by that dog truly does offer us a way into our own souls. No better put than in the exquisite words of Anatole France,

"Until one has loved an animal, a part of one's soul remains unawakened."

Dhalia running free on the Granite Dells.

Dhalia running up on the Granite Dells, near Payson, AZ.

Chapter 26

MESSAGES FROM THE NIGHT

A story.[72]

"Jean, where's Dhalia?"

"Don't know. She was here moments ago."

"Jeannie, you take the other dogs back to the car and I'll go and scout around for her. Oh, and you better put Pharaoh on his leash, otherwise you know he'll follow me."

"Paul, don't worry, Dhalia's always chasing scents; bet she beats us back to the car. Especially as it's going to be dark soon."

Nonetheless, I started back down the dusty, dirt road, the last rays of the sun pink on the high, tumbled cliffs of granite. This high rocky forest plateau, known as the Granite Dells, just three miles from our home on the outskirts of Payson, Arizona, made perfect dog-walking country and rarely did we miss an afternoon out here. However this afternoon, for reasons I was unclear about, we had left home much later than usual.

There was no sign of Dhalia ahead on the road, so I struck off left, hoping she was somewhere up amongst the trees and the high boulders. Soon I reached the first crest, panting hard in the thin air. Behind me, across the breathtaking landscape, was a magnificent sight of the setting sun just dipping beneath faraway mountain ridges. Then, quite suddenly, in the midst of my brief pause admiring this perfect evening, a sound echoed around the cliffs, the sound of a dog barking. I bet my life on that being Dhalia. Just as quickly the barking stopped.

The barking started up again, barking that suggested Dhalia was hunting something. The sound came from an area of boulders way up above the pine trees on the other side of the small valley ahead of me.

Perhaps, Dhalia had trapped herself. More likely, I reflected, swept up in the evening scents of the wilderness, Dhalia had temporarily reverted back to the wild, hunting dog she had been all those years ago. That homeless Mexican street dog who in 2005 had tentatively turned away from scavenging in a pile of rubbish in a dirty Mexican town and shyly approached Jean. Jean had named her Dhalia.

I set off down to the valley floor and after fifteen minutes of hard climbing had reached the high boulders on the other side. I waited to get my breath back and whistled, then called "Dhalia. Dhalia. Come on, there's a good girl."

Thank goodness, Dhalia was such a sweet, obedient dog.

I anticipated the sound of dog feet scampering through rough undergrowth. But no sound came.

I listened: no sounds, no more barking. Now where had she gone? Perhaps past these boulders down into the steep ravine beyond me, the one so densely forested with pine trees.

With daylight practically gone, I needed to find Dhalia very soon. I plunged down the slope, pushing through tree branches that whipped across my face, then fell heavily as a foot found empty space instead of the anticipated firm ground.

I cursed, picked myself up, and paused. That fall had a message: the madness of continuing this search in the near dark. The terrain made very rough going, even in daylight. At night, the boulders and plunging ravines would guarantee a busted body, at best. Plus, I ruefully admitted, I didn't have a clue about finding my way back to the road from wherever I now was.

The unavoidable truth smacked me full in the face. I would be spending this night alone in the high, open forest.

It had one hell of a very scary dimension. I forced myself not to dwell on just how scary it all felt. I needed to stay busy, find some way of keeping warm; last night at home it had dropped to within a few degrees of freezing. I looked around, seeing a possible solution. I broke a small branch off a nearby mesquite tree and made a crude brush with which I swept up the fallen pine needles I saw everywhere about me. Soon I had a stack sufficient to cover me, or so I hoped.

Thank goodness that when Jeannie and I had decided to give four of our dogs this late afternoon walk, I had jeans on and was wearing a long-sleeved shirt, a pullover thrown over my shoulders. It didn't make Dhalia's antics any less frustrating, but I probably wasn't going to freeze to death.

The air temperature sank as if connected with the last rays of the sun. My confidence sank in harmony with the falling temperature. I lay down, shuffled about, swept the pine needles across my body, tried to find a position that carried some illusion of comfort. No matter the position, I couldn't silence my mind. No way to silence the screaming in my head, this deep, primeval fear of the dark forest about me, imagination already running away with visions of hostile night creatures, large and small, watching me, smelling me, biding their time.

Perhaps I might sleep for a while? A moment later the absurdity of that last thought hit me. Caused me to utter aloud, "You stupid sod. There's no way you're going to sleep through this."

My words echoed off unseen cliffs in the darkness, reinforcing my sense of isolation. I was very frightened. Why? Where in my psyche did that come from? I had spent many nights alone at sea without a problem, a thousand miles from shore. Then, of course, I knew my location and always had a radio link to the outside world. But being lost in this dark, lonely forest touched something very deep inside of me.

Suddenly, I started shivering. The slightest movement caused the needles to slip from me and the cold night air began to penetrate my body. I pondered about how cold it might get and, by extension, thanked my lucky stars that it was early October not, say, mid-December. So far, not too cold, but soon the fear rather than the temperature started to devour me. What stupid fool said, "Nothing to fear but fear itself"? My plan to sleep under pine needles, fear or no fear, had failed. I couldn't get warm. I had to move.

Looking around, I saw in the gloom a few yards away an enormous boulder, like some giant, black shadow. No details, just this huge outline etched against the night. I carefully raised myself, felt the remaining needles fall away, and gingerly shuffled across to the dark rock. I half-expected something to bite my extended hand as I explored the surface, as I ran my hand down towards the unseen ground. Miracle of miracles, the granite gently emitted the warmth absorbed from the day's sun. I slowly settled myself to the ground, eased my back against the rock-face and pulled my knees up to my chest. I felt so much less vulnerable than when I had been flat out on the forest floor. I let out a long sigh, then burst into tears, huge heart-rending sobs coming from somewhere deep within me.

Gradually the tears washed away my fear, restored a calmer part of my brain. That calmer brain brought the realisation that I hadn't considered, well not until now, what Jeannie must be going through. At least I knew I was alive and well. Jeannie, not knowing, would be in despair. I bet she would remember that time when out walking here in the Dells we had lost little Poppy, an adorable ten-pound poodle mix, never to be found again despite ages spent combing the area, calling out her name. A year later and Jeannie still said from time to time, "I so miss Poppy". There was no question that I had to get through this in one piece, mentally as much as physically.

Presumably, Jeannie would have called 911 and been connected to the local search-and-rescue unit. Would they search for me in the dark? I thought that unlikely.

Thinking about Jeannie further eased my state of mind and the shivering stopped. Thank goodness for that. I fought to retain this new perspective. I would make it through, even treasure this night under the sky, this wonderful, awesome, night sky. Even the many pine tree crowns that soared way up above me couldn't mask a sky that just glittered with starlight. Payson, at 5,000 feet, had many beautifully clear skies and tonight offered a magical example of that.

Frequently during my life, the night skies had spoken to me, presented a reminder of the continuum of the universe. On this night, however, I felt more humbled by the hundred million stars surrounding me than ever before.

Time slipped by, my watch in darkness. However, above my head was that vast stellar clock. I scanned the heavens, seeking out familiar pinpoints of light, companions over so much of my lifetime. Ah, there: The Big Dipper, Ursa Major, and, yes, there's Polaris the North Pole. Great! Now the rotation of the planet became my watch, The Big Dipper sliding around Polaris, fifteen degrees for each hour.

What a situation I had got myself into. As with other challenging times in my life, lost in the Australian bush, at sea hunkering down through a severe storm, never a choice other than to work it out. I felt a gush of emotion from the release this changed perspective gave me.

Far away, a group of coyotes started up a howl. What a timeless sound it seemed to represent. How long had coyotes been on the planet? I sank into those inner places of the mind noting how the intense darkness raised some deep thoughts. What if this night heralded the end of my life, the last few hours of the life of Paul Handover? What parting message would I give to those I loved?

Jeannie would know beyond any doubt how much I had adored her, how her love had created an emotional paradise for me beyond measure. But my son and daughter, dear Alex and Maija? Oh, the complexities I had created in their lives by leaving their mother so many years ago. I knew that they still harboured raw edges, and quite reasonably so. I still possessed raw edges from my father's death, way back in 1956. That sudden death, just five days before Christmas and so soon after I had turned twelve, had fed a life-long feeling of emotional rejection. The feeling that had lasted for fifty-one years until, coincidentally also five days before Christmas, when on December 20, 2007, I had met Jean.

My thoughts returned to Alex and Maija. Did they know, without a scintilla of doubt, that I loved them? Maybe my thoughts would find them. Romantic nonsense? Who knows? Dogs had the ability to read the minds of humans, often from far out of visual range. I knew Pharaoh, my devoted German Shepherd, skilfully read my mind.

I struggled to remember that saying from James Thurber. What was it now? Something about men striving to understand themselves before they die. Would that be my parting message for Alex and Maija? Blast. I wished I could remember stuff more clearly these days and let go of worrying about the quote. Perhaps my subconscious might carry the memory back to me.

I looked back up into the heavens. The Big Dipper indicated at least an hour had slipped by. Gracious, what a sky in which to lose one's mind. Lost in this great cathedral of stars. Then, as if through some stirring of my consciousness, that Thurber saying did come back to me: "*All men should strive to learn before they die, what they are running from, and to, and why.*"

I reflected on those who, incarcerated in solitary confinement, had their minds play many tricks, especially when it came to gauging time. What a bizarre oddment of information, where had that come from? Possibly because I hadn't a clue about my present time. It felt later than 11 p.m. and earlier than 4 a.m., but any closer guess seemed impossible. Nevertheless, from out of those terrible, heart-wrenching hours of being alone I had found calm, had found a peace within. I slept.

Suddenly, a sound slammed me awake. Something out there in the dark had made a sound, caused my whole body to become totally alert, every nerve straining to recognise what it might be. It sounded like animal feet moving through the autumn fall of dead leaves. I prayed it wasn't a mountain lion. Surely such a wild cat preparing to attack would be silent. Now the unknown creature had definitely paused, no sound, just me knowing that out there something waited. Now what? The creature had started sniffing. I hoped not a wild pig. Javelinas, those pig-like creatures that always moved in groups, could make trouble, they had no qualms at attacking a decent-sized dog.

Poised to run, I considered rising to my feet but chose to stay still and closed my right hand around a small rock. The sniffing stopped. Nothing now, save the sound of my rapidly beating heart. I sensed, sensed strongly, the creature looking at me. It seemed very close, ten or twenty feet away. The adrenalin hammered through my veins.

I tried to focus on the spot where I sensed the animal waited; waiting for what? I pushed that idea out of my head. My ears then picked up a weird, bizarre sound. Surely not? Had I lost my senses? It sounded like a dog wagging its tail: flap, flap, flapping against something such as a tree trunk.

A dog? If a dog, it had to be Dhalia.

Then came that small, shy bark. A bark I knew so well. Oh, wow, it is Dhalia. I softly called, "Dhalia, Dhalia, come here, there's a good girl."

With a quick rustle of feet, Dhalia leapt on me, tail wagging furiously, her head quickly burrowing into my body warmth. I hugged her and, once more, tears poured down my face. Despite the darkness, I could see her perfectly in my mind, her tight, short-haired coat of light-brown hair, her aquiline face, her bright inquisitive eyes, and those wonderful head-dominating ears. Lovely large ears that seemed to listen to the world. A shy, loving dog when Jean had rescued her in 2005 and all these years later still a shy, loving dog.

Dhalia licked my tears, her gentle tongue soft and sweet on my skin. I shuffled more onto my back, which allowed her to curl up on my chest, still enveloped by my arms. My mind drifted away to an era a long time ago, back to an earlier ancient man, likewise with arms wrapped around his dog under a dome of stars, bonded in a thousand mysterious ways.

The morning sun arrived as imperceptibly as an angel's sigh. Dhalia sensed the dawn before I did, brought me out of my dreams by the slight stirring of her warm, gentle body.

Yes, there it came, the end of this night. The ancient sun galloping towards us across ancient lands; another beat of the planet's heart. Dhalia slid off my chest, stretched herself from nose to tail, yawned, and looked at me, as much to say time to go home. I could just make out the face of my watch: 4.55 a.m. I, too, raised myself, slapped my arms around my body to get some circulation going. The cold air stung my face, yet it couldn't even scratch my inner warmth, the gift from the loving bond that Dhalia and I were sharing.

We set off and quickly crested the first ridge. Ahead, about a mile away, we saw the forest road busy with arriving search-and-rescue trucks. I noticed Jean's Dodge parked ahead of the trucks and instinctively knew she and Pharaoh had already disappeared into the forest, Pharaoh leading the way to us.

We set off down the slope, Dhalia's tail wagging with unbounded excitement, me ready to start shouting for attention from the next ridge. We were about to wade through a small stream when, across from us, Pharaoh raced out of the trees. He tore through the water, barking at the top of his voice in clear dog speak *I've found them, they're here, they're safe*. I crouched down to receive my second huge face lick in fewer than six hours.

Later, once safely home, it came to me. When we had set off in that early morning light, Dhalia had stayed pinned to me. So unusual for her not to run off. Let's face it, that's what got us into the mess in the first place. No, Dhalia had stayed with me as if she had known that during that long, dark night, it had been me who had been the lost soul.

The message from the night, as clear as the rays of this new day's sun, the message to pass to all those I love. If you don't get lost, there's a chance you might never be found.

Jean walking Pharoah on the Granite Dells, AZ.

Chapter 27

EMBRACING DEATH

I'm sure that the human psyche lives in a bubble of self-delusion. Not always and not extremely so, of course, for if the level of delusion were abnormal then we couldn't function properly as social beings. However, just be still for a moment and reflect on the ways that you shelter from reality under certain circumstances. In directing that question to you, dear reader, trust me I do not exclude myself.

There are times when going beyond the self, going out of oneself, is the only way to see the reality of who we are and the world around us; for us to be able to brush away our delusions. Perfectly mirrored by the words of Aldous Huxley: "Experience is not what happens to you; it's what you do with what happens to you." Wise words, indeed.

But these fine expressions have one grinding, searing fault. They do not assume the end of a person's life. I am speaking of death, of the inevitability of our death. That largely unspoken question that Sharon Salzberg writes in her book *Faith: Trusting Your Own Deepest Experience*[73] "What does it mean to be born in a human body, vulnerable and helpless, then to grow old, get sick and die, whether we like it or not?"

Anyone who has lost a loved one knows that it is tough, incredibly tough. It is full of pain and anguish amidst a great churning of emotions, all going on in a very deep-seated and personal manner. That's the perspective from the loved ones left behind with more life ahead of them. But if one thinks of it in reverse, turns it on its head so to speak, what are our fundamental wishes with respect to those we love: what we would want to leave behind us when *we* die?

Fond memories, naturally, but they wouldn't be our fundamental wishes. Surely, our fundamental wishes would be that our death does not leave pain and anguish in the hearts and minds of those left behind. That it doesn't leave a pain that cannot be dealt with in a healthy way. Our wishes would be that those whom we loved and who loved us might embrace their loss, and then carry on living their lives.

Anyone who has loved a dog has most likely been intimately involved in the end of that dog's life. It is, to my mind, the ultimate lesson that dogs offer us: how to be at peace when we die and how to leave that peace blowing like a gentle breeze through the hearts of all the people who loved us. (And to underline that ultimate lesson, we knew dear Lilly, who had been loved by Jean for more than sixteen years, was very close to the end of her life towards the end of me writing this book. On Sunday, 23 August 2015, Jim Goodbrod, DVM, euthanised Lilly, aged 17, while she lay quietly asleep in Jean's arms. Later that same day I finished the final edit of this manuscript.)

Our beloved dogs have much shorter life spans than we do, thus almost everyone who has loved a dog will have had to say goodbye to that gorgeous friend of theirs at some point in their lives. Very sadly, perhaps, saying goodbye to more than one loved dog.

I see the most precious of parallels in the tragic death of a loved dog and our own deaths. The parallel between coping with our grief at the loss of our loved dogs, and reaching out to *our* loved ones so that they might cope with their grief at losing us.

Put another way: knowing what to expect in emotional terms at the loss of our loved dog is helpful, very much so, to us helping our loved ones when it comes our time to die.

I discovered that there are five stages of mourning,[74] of dealing with our grief, when we lose our beloved dog: denial, anger, guilt, depression, and acceptance.

I then compared those stages to the five stages of mourning[75] for the death of a human. Those five stages are: denial and isolation, anger, bargaining, depression, and acceptance. They apply whether we are dealing with the knowledge that we are dying, or the impending death of a person who is very emotionally close to us.

The parallels between the stages of mourning for the death of a beloved dog and the mourning for a human are perfect. Whether it is the impending or actual death of a loved dog or a loved person, the similarities between embracing the loss of the dog or the person are powerfully obvious. So, too, are the many different ways each of us embraces the death of that loved dog or loved person. Let me expand on that.

Namely, that each of us will experience each stage of mourning at varying levels of intensity, for varying lengths of time, and sometimes in a different order. Some of the stages might converge and overlap one another but however you experience the mourning, it is incredibly important to remember that your feelings are completely normal.

As the website of the American Animal Hospital Association points out, on its webpage titled "Life after Dog",[76] "we almost always outlive our beloved companions. Learning to live with loss is an essential part of life."

That association webpage implores us to honour our emotions, to honour the memory of our dog, and critically, when a child is involved, to help that child cope with the loss of their loved dog. Helping to ease the pain for all, young and old, through learning to cope with the loss. Easing the pain through changing one's schedule, moving furnishings around to help distance the memory of your dog's favourite sleeping spots, or creating a memorial in one form or another, even writing a letter to your dog in which you describe all the feelings you have for your recently departed, loving friend.

Chapter 28

CONCLUSION

I was born in 1944. I am therefore the wrong side of seventy years old. I was born an Englishman and, according to life expectancy tables from 2012, a male Englishman's life expectancy is 79.5 years. I am living happily in the United States and, according to those same tables, a male American's life expectancy is 77.4 years. My mother is alive and an amazingly fit and healthy 95 years old, as of August 2015. My father died at the age of 56 in 1956, just five days before Christmas.

None of us knows how the end will come, just that one day it will. The inevitability of death.

All that I do know is that loving our dogs, welcoming all the wonderful qualities that our dogs possess, striving always to live peacefully in the present, just as our dogs do, and, ultimately, as with our faithful companions, taking that last breath in the knowledge that ours was a beautiful life, is what learning from dogs is all about.

I'm making the assumption that if you have reached this last chapter, the conclusion, then you have read the preceding chapters. (And if that isn't the case, then please keep that to yourself!)

The author of any book must suffer from the disadvantage of never reading it with fresh eyes. As I reflect on what would be the appropriate message to convey in this conclusion, from this first-time author, my mind drifts away from that reflection to somewhere else. Aided and abetted by the sight out of my window, no doubt. It is approaching 6 p.m. on a Saturday evening in August. There has been a hazy sky today due to smoke from a couple of forest fires, luckily some distance away. The air is still, the setting sun, a deeper red than normal thanks to that smoke in the sky, is dipping below the tops of tall trees, pine, fir and oak, close by the house.

We live in a rural part of southern Oregon and frequently the interchange between sun and shade, the visual harmony of grass fields, forest, and mountain range, delivers a profound sense of timelessness. When so emotionally moved it seems almost destructive, in a psychological way, to return my thoughts to what motivated me to write this book in the first place.

Yet there is a link between these two states of mind. For no-one knows what is around the corner. Nature will always have the last word regarding her natural world, to which we humans are so intricately linked. Standing alongside and respecting nature as the future comes to us will be so much wiser than pushing back against nature, and ultimately failing, trying to "convert" nature to some form of materialistic human resource. Because that route will only return those of us who survive to a life of hunting and gathering. Which, so many thousands of years previously, is where early dogs started humankind on the long journey leading to now.

Dogs have been the making of humans and a viable future for humankind on this beautiful planet depends on us never forgetting this oldest relationship of all, the one between dog and human.

This book opened with the words of Dr. Jim Goodbrod, DVM. I think it should close with me repeating the quotation that Dr. Jim used in his Foreword, that one from Suzanne Clothier's book *Bones Would Rain from the Sky: Deepening Our Relationships with Dogs*:

> There is a cycle of love and death that shapes the lives of those who choose to travel in the company of animals. It is a cycle unlike any other. To those who have never lived through its turnings or walked its rocky path, our willingness to give our hearts with full knowledge that they will be broken seems incomprehensible. Only we know how small a price we pay for what we receive; our grief, no matter how powerful it may be, is an insufficient measure of the joy we have been given.

Appendix 1

The Wolf Who Stayed: A Domestication That Went Both Ways

Mark Derr[77]

That the dog is descended from the wolf—or more precisely, the wolf who stayed—is by now an accepted fact of evolution and history. But that fact is about all that is agreed to among the people who attempt to answer fundamental questions about the origins of the dog—specifically, the who, where, when, how and why of domestication.

Dates range from the dog's earliest appearance in the archaeological record around 14,000 years ago to the earliest estimated time for its genetic sidestep from wolves around 135,000 years ago. Did the dog emerge in Central Europe, as the archaeological record suggests, or in East Asia, where the genetic evidence points? Were they tame wolves whose offspring over time became homebodies, or scavenging wolves whose love of human waste made them increasingly tame and submissive enough to insinuate themselves into human hearts? Or did humans learn to follow, herd and hunt big game from wolves and in so doing, enter into a complex dance of co-evolution?

Despite the adamancy of adherents to specific positions, the data are too incomplete, too subject to wildly different interpretations; some of the theories themselves too vague; and the physical evidence too sparse to say with certainty what happened. Nonetheless, some models, and not necessarily the most popular and current ones, more clearly fit what is known about dogs and wolves and humans than others. It is a field in high flux, due in no small measure to the full sequencing of the dog genome. But were I a bettor, I would wager that the winning view, the more-or-less historically correct one, shows that the dog is the result of the interaction of wolves and ancient humans rather than a self-invention by wolves or a "conquest" by humans.

Our views of the dog are integrally bound to the answers to these questions, and, for better or worse, those views help shape the way we approach our own and other dogs. It is difficult, for example, to treat as a valued companion a "social parasite" or, literally, a "shit-eater." To argue that different breeds or types of dogs represent arrested stages of wolf development both physically and behaviorally is not only to confuse, biologically, description with prescription but also to overlook the dog's unique behavioral adaptations to life with humans. Thus, according to some studies, the dog has developed barking, a little-used wolf talent, into a fairly sophisticated form of communication, but a person who finds barking the noise of a neotenic wolf is unlikely to hear what is being conveyed. "The dog is everywhere what society makes him," Charles Dudley Warner wrote in *Harper's New Monthly Magazine* in 1896. His words still hold true.

Since the dog is both a cultural and a biological creation, it is worth noting here that these opposing views of the dog's origin echo the old theory that the sniveling, slinking pariah dogs and their like, "southern breeds" derived from jackals, while "northern breeds" Spitz-like dogs and Huskies descended directly from the wolf. Darwin thought as much, so did the pioneering ethologist Konrad Lorenz until late in his life, when he accepted that the wolf was the sole progenitor of the dog. In the theories of Raymond Coppinger and others, and I think this transference is unconscious, the scavenging jackal becomes a camp-following, offal-eating, self-domesticating weenie of a tame wolf. In turn, those wolves become the ur-dog, still manifest in the pariahs of India and Asia, from which the dog we know is said to have emerged. It's a tidy, convenient, unprovable story that has an element of truth, dogs are accomplished scavengers, but beyond that, it is the jackal theory with a tattered new coat. In dropping humans from the process, the scavenging, self-domesticating wolf theory ignores the archaeological record and other crucial facts that undercut it.

Fossils found at Zhoukoudian, China, have suggested to archaeologists such as Stanley Olsen, author of Origins of the Domestic Dog, that wolves and Homo erectus were at least working the same terrain as early as 500,000 years ago. The remains of wolves and Homo erectus dating to around 300,000 years ago have also been found in association with each other at Boxgrove in Kent, England, and from 150,000 years ago at Lauzerte in the south of France. It seems more likely that this omnivorous biped, with its tools and weapons, lived and hunted in proximity to that consummate social hunter, the wolf, through much of Eurasia, than that their bones simply fell into select caves together. Who scavenged from whom, we cannot say.

Wolves were far more numerous then than now, and they adapted to a wide range of habitats and prey. On the Eurasian steppes, wolves learned to follow herds of ungulates, in effect, to herd them. Meriwether Lewis observed the same behavior during his journey across North America in the opening years of the 19th century; he referred to wolves that watched over herds of bison on the Plains as the bisons' "shepherds." Of course, those "shepherds" liked it when human hunters attacked a herd because they killed many more animals than the wolves, and although the humans carried off the prime cuts, they left plenty behind.

Ethologists Wolfgang M. Schleidt and Michael D. Shalter refer to wolves as the first pastoralists in "Co-evolution of Humans and Canids," their 2003 paper in the journal Cognition and Evolution. Early humans, they argue, learned to hunt and herd big game from those wolves; thus, the dog emerged from mutual cooperation between wolves and early humans, possibly including Neanderthal. There is no evidence yet of Neanderthal having tame wolves, much less dogs, but the larger point is that when modern humans arrived on the scene, they found wolves already tending their herds, and they immediately began to learn from them. That was long before humans began, in some parts of the world, to settle into more permanent villages, some 12,000 to 20,000 or 25,000 years ago.

Schleidt and Shalter based their model on wolf behavior and on genetic studies that have consistently shown that dogs and wolves diverged between 40,000 and 135,000 years ago. The first of those studies emerged from the lab of Robert K. Wayne, an evolutionary biologist at the University of California at Los Angeles who had already made headlines by showing definitively that the dog descended from the wolf alone. In a paper appearing in the June 13, 1997, issue of Science, Wayne and his collaborators said that dogs could have originated around 135,000 years ago in as many as four different places. They also argued that genetic exchanges between wolves and dogs continued as they do to this day, albeit in an age during which dogs have become ubiquitous and wolves imperiled.

Since that paper appeared, the dog genome has been fully sequenced and provides a time frame for domestication of 9,000 generations, which the authors of a paper on the sequencing in the December 8, 2005, issue of Nature pegged at 27,000 years. But except for that, subsequent studies of mitochondrial DNA, which is most commonly used to date species divergence, have pointed to a time frame of 40,000 to 135,000, with 40,000 to 50,000 years ago looking like the consensus date.

Most of this work has been conducted in Wayne's lab; in the Uppsala University lab of Carles Vilà, his former student and the lead researcher on the 1997 paper; and in the lab of Peter Savolainen of the Royal Institute of Technology, Stockholm, another collaborator on the original paper.

A signal problem with the early date is that it doesn't appear to match the archaeological record. The dog is not only behaviorally but also morphologically different from the wolf, and such an animal first appears in the fossil record around 14,000 years ago in Bonn-Oberkassel, Germany. Archaeologists nearly universally peg the origin of the dog to that time.

Wayne, Vilà and their supporters have suggested from the start that behavioral change could predate morphological change, which would have occurred when humans began to create permanent settlements, thereby cutting, or at least reducing, their wolf-dogs' contact with wild wolves. People might also have begun attempting to influence the appearance of their dogs at this point.

But those Germans get in the way again. Bonn-Oberkassel, site of the consensus first fossil dog, is not a permanent settlement.

Trying to square genetic and archaeological dates, Peter Savolainen resurveyed the mitochondrial DNA of dogs and wolves, recalibrated the molecular clock and proposed in a paper in Science, November 22, 2002, that the dog originated in East Asia 15,000 to 40,000 years ago. It was a good try, but now it appears that his "40,000 years ago" date was more accurate. Also, the earliest known dog appears in Germany, not East Asia, a region to which other genetic evidence points as well.

In many ways, the dispute over dates and places is just a precursor for the debate over how that happened. Archaeologists and evolutionary biologists who want the first dogs to look like dogs have tended to argue that the transition is a result of a biological phenomenon called "paedomorphosis." That basically means that the animal's physical development is delayed relative to its sexual maturation. It produces dogs with more domed heads; shorter, broader muzzles; and overall reduced size and slighter build than a wolf. Accelerated physical development relative to sexual maturation (hypermorphosis), on the other hand, produces dogs larger than the progenitor wolf.

When maturation is stopped early enough, the resulting animal is said to resemble a "neotenic," or perpetually juvenilized, wolf. Coppinger and others have carried the argument further to argue that behaviorally, the dog resembles a neotenic wolf, with some breeds being more immature or less developed than others. There is general agreement that, beginning in the late 19th century when the dog began to move into the city as a pet, breeders sought to soften and humanize the appearance of some breeds to make them look like perpetual puppies. But beyond that, it is more correct to view the dog as an entity different from the wolf.

Currently, many researchers like to invoke an experiment in domestication launched in 1959 at the Institute of Cytology and Genetics in Novosibirsk, Siberia, by Dmitry Belyaev and continued after his death by Lyudmila Trut and her colleagues. Belyaev selectively bred foxes for "tameness" alone, defined as their level of friendliness toward people. He ended up with foxes that resembled dogs. A number of them had floppy ears, piebald coats, curly tails and a habit of submissively seeking attention from their human handlers with whines, whimpers and licks. (I wouldn't want such a dog.)

Anthropologist Brian Hare tested the tame foxes in 2004 and found that they, like dogs, had the capacity to follow a human's gaze, something wolves and wild foxes, not to mention chimpanzees, won't do.

A number of researchers have embraced these tame foxes as a template for dog domestication. While they doubtless cast insight on the problem, I doubt that they will answer all questions. Arguments by analogy are suspect science and should be even more so in this case, since the selection criteria for these foxes were also against aggression, hardly the case for dogs, and foxes clearly are not wolves.

That said, the experiment does appear to confirm that selective breeding for behavior alone can also produce morphological changes similar to what the wolf experienced in becoming a dog.

Coppinger has invoked the fox experiment to support his theory that wolves that became dogs self-domesticated. As humans in some areas moved into permanent settlements, their refuse heaps became feeding grounds for wolves who were tame enough, or least-frightened enough, to feed near humans. Subsequent generations became more tame, and people began to allow them to wander their camps, eating feces, hunting rodents. From that group, people took some animals for food. Then, when the animals were thoroughly self-tamed, people began to train them to more wolfish behaviors, like hunting.

What he and others overlook in citing the fox experiment is that those animals were subjected to intense artificial selection by people. They also ignore the fact that the first dog appears in a seasonal camp, not a permanent settlement.

In their book, *Dogs*, Coppinger and his wife, Lorna, argue that these early protodogs would have resembled the ownerless dogs of Pemba Island, a remote part of the Zanzibar archipelago. As a model, Pemba suffers numerous problems, as does Coppinger's theory. It is an Islamic island, and Islam has scarce place for dogs, believing them filthy, largely because they scavenge and eat excrement.

Beyond that, Pemba was a wealthy island in the 18th and 19th centuries due to its clove plantations, which were worked by African slaves and overseen by Arabs. The plantations have long since fallen into disrepair, on an island populated by the descendants of free slaves, where poverty is the rule. Attempting to read the past by looking at the present is a well recognized form of historical fallacy. It can't be done, especially in a place where there is no strong cultural tradition.

Elsewhere in the developing world, free-ranging dogs are often more than scavengers or food. Some are fed; they protect territories or vendors' carts. A few might be taken in, but, again, these dogs must be studied and understood in their current context and then placed in a broader historical context, if possible.

Moreover, Coppinger ignores the entire tradition of dogs and people in Europe, Japan and Korea, wherever dogs were employed from an apparently early date for a purpose, including companionship and ritual. Archaeologist Darcy F. Morey clearly demonstrated in the February 2006 issue of The Journal of Archaeological Science that people have been burying dogs and treating them with reverence and respect from the beginning, hardly the fate of scavengers.

People will argue, but I think the question of whether the dog is a juvenilized wolf is best answered with this observation: The dog follows human gaze, according to Hare, and is so attentive to people that it can imitate them, according to Vilmos Csányi, and it does so from an early age. No wolf of any age can replicate that basic behavior. It is far better to look at the dog as a differently developed wolf than as a developmentally retarded wolf.

Similarly, until shown otherwise, it seems more accurate to view domestication as a dynamic process involving wolves and people. At a time when the boundaries between human and wild were much more porous than now, people doubtless took in animals, especially young animals of all kinds, especially wolf pups, since in many places, they were hunting the same game and perhaps scavenging from each other.

As those pups matured, they returned to the wild to breed, with the naturally tamest among them denning close to the camp where they had been raised and, yes, could scavenge. Over the past year, researchers have shown that the area of the brain known as the amygdala is quite active when "fear of the other" begins to develop. In 2004, a team of researchers from Uppsala University, including Vilà, reported in the journal Molecular Brain Research on changes they had found in gene expression in the frontal lobe, hypothalamus and amygdala of wolves, coyotes and dogs. More than 40 years ago, J.P. Scott and John L. Fuller showed that the dog pup had a lengthened socialization period before fear of the other set in, compared with the wolf pup.

No one knows how fast the change happened, but in some places, tame wolves, dogs, resulted from this process. They provided territorial defense, helped with hunting (which they do well), scavenged, and were valued for companionship and utility. Some could be trained to carry packs. That early dog probably remained nearly indistinguishable from the wolf except in places where their gene pool became limited by virtue of some isolating event. The smaller gene pool forced inbreeding that, along with changing environmental conditions, somehow "destabilized" the genome.

Vilà and two colleagues suggested in an article published online on June 29, 2006, in Genome Research, that domestication relaxed "selective constraint" on the dog's mitochondrial genome, and if that relaxation extended to the whole genome, as it appeared to, "it could have facilitated the generation of novel functional genetic diversity."

European and North American breeders have taken full advantage of that or some other mechanism to create the most morphologically diverse mammal around. But other cultures did not follow that path.

There are other theories afloat in what is an exciting time for people who study dogs. But the one that succeeds will reflect the dynamic relationship between human and dog.

This article first appeared in *The Bark*, Issue 38 (Sep/Oct 2006). Mark Derr is the author of *A Dog's History of America*, *Dog's Best Friend*, *The Frontiersman: The Real Life and Many Legends of Davy Crockett*, *Some Kind of Paradise*, and *How the Dog Became the Dog*, and numerous articles on science, environment and transportation. He blogs for *Psychology Today*.

Ben showing trust and forgiveness to the author.

Appendix 2

A Horse Called Ben
A Lesson in Forgiveness

Jean and I, and all our dogs and cats, moved to a rural property in Oregon in October 2012. Previous owners of the property must have been horse people, for there were five stables plus a hayloft. Therefore, once we had settled ourselves in, we decided that we had sufficient acres of pasture and the necessary facilities to have a horse. As a young lady, Jean had been a keen horse rider.

We were put in touch with an all-breed horse rescue and rehabilitation centre, Strawberry Mountain Mustangs, near Roseburg, Oregon, under the care and ownership of a Darla Clark. The centre was about an hour north from where we lived and in due course we arranged to visit.

During that visit we took a liking to a 15-year-old gelding. His name was Ranger and he had been found abandoned in the Ochoco National Forest in central Oregon, subsequently arriving at Strawberry Mountain. Ranger had a delightful temperament plus his age was a bonus as both Jean and I preferred not to be taking on a horse that might outlive us!

We returned home and started organising the stable area and checking that the acres of grassland available for Ranger were securely fenced. However, just a week before Ranger was due to be transported down to us, there was a telephone call from Darla asking us whether we might consider taking *two* horses.

It transpired that this second horse, named Ben, was a younger horse that had also been "rescued" (on the orders of Roseburg's local sheriff) because of Ben at the time being in private ownership. It turned out that Ben had been subjected to starvation, to regular beatings, and even worse than that, there was evidence that he had been repeatedly shot in the chest with an air gun. The sheriff's office legally took away Ben and placed him under the care of Darla. Apparently, Ben was deeply traumatised, as one might imagine, and quickly formed a close relationship with Ranger. Darla was in no doubt that Ben's relationship with Ranger was an essential part of his journey of returning to a healthy, confident horse.

There was no cause for us saying "no" to take both Ben and Ranger, for we had the stabling and grassland to accommodate a number of horses. Darla did explain that Ben presently was a very wary horse, especially nervous of men, and that I should never make any sudden movements around him; mostly for my own safety.

A week later, Darla came to us with Ranger and Ben.

Unlike Jean who had owned and ridden horses in her younger days, I hardly knew the front from the back of a horse. I decided to approach Ben as I would a new rescue dog.

In fewer than three weeks, Ben had recovered sufficient trust of men to allow me to stroke his neck. Three months later, I could put my face against Ben's muzzle and at the same time stroke the area on his chest that was covered in gun pellet scars.

If only humans could learn to forgive in such a manner. Many people experiencing these levels of cruelty would harbour anger and distrust in their hearts forever.

Author and Ben revealing how love and kindness can restore trust in animals.

About the Author

Paul Handover is a child of the post-war era in Great Britain having been born in London six months before the end of World War II. After a rather shaky attempt at being educated, including two years studying for a diploma in electrical engineering, Paul's first job was as a commercial apprentice at the British Aircraft Corporation in Stevenage, Hertfordshire. He then joined the sales desk of British Visqueen, a polythene film and products manufacturer and part of ICI Plastics Division. In 1968, Paul emigrated to Australia and became part of the sales team at ICI's Inorganic Chemicals Division in Sydney.

The author out walking on Dartmoor, southwest England, January 2008.

While living in Sydney, a friendship between Paul and Norman Nicholls, a professional photographer, led to Paul being appointed as a freelance features writer for the Finnish magazine Koti Posti (both Paul and Norman were married to Finnish women), resulting in Paul travelling all over Australia writing about Finns and their lives.

After returning to the United Kingdom in 1970, Paul signed up with IBM's Office Products division as a field salesman and spent eight very happy years with the company. But the lure of starting a business was too strong and in 1978 he left IBM to start his own company, Dataview Ltd. Dataview became the eighth Commodore PC dealer in the United Kingdom and later the eighth IBM PC dealer. The company also became famous for producing and globally distributing a British-made word processing program: Wordcraft.

During this exciting period, Paul became much more aware of the importance of marketing strategy and the raft of competencies that deliver entrepreneurial success. After Dataview was sold, he used these experiences to act successfully as a sales-and-marketing consultant for small and large organisations. Paul has also been a visiting teacher at both English and French business schools and continues to provide coaching and mentoring to entrepreneurs both sides of the pond.

In July 2009 Paul started writing a blog named "Learning from Dogs", that he still maintains on a daily basis.

Paul now lives in Merlin, Oregon, United States, with his wife, Jean, also born a Londoner.

ENDNOTES

1. J has requested that his identity not be revealed in the book.

2. "The Origin of Dogs," *Scientific American*, (August 20, 2009).

3. https://en.wikipedia.org/wiki/List_of_archaeological_periods.

4. http://www.infoplease.com/encyclopedia/society/mesolithic-period.html, and others.

5. Marc Bekoff, Ph.D. professor emeritus of ecology and evolutionary biology at the University of Colorado, Boulder.

6. http://en.wikipedia.org/wiki/Mark_Derr.

7. *The Bark*, Issue 38 (Sep/Oct 2006.)

8. http://www.nbcnews.com/id/27240370/ns/technology_and_science-science/t/worlds-first-dog-lived-years-ago-ate-big/.

9. A paleontologist at the Royal Belgian Institute of Natural Sciences.

10. See http://news.discovery.com/animals/pets/prehistoric-dog-lovers-profiled-130521.htm.

11. Associate Professor in the department of anthropology, University of Alberta.

12. Published by Bloomsbury.

13. I am indebted to Suzann Reeves, sister of Dan Gomez, for not only steering me towards taking on a German Shepherd breed but for also coming up with the name: Pharaoh. Well done, Su.

14. George B. Johnson, PhD, is a professor of biology at Washington University in St. Louis and a professor of genetics at the university's school of medicine.

15. My relationship was terminated when I became a resident of the United Staes in 2011.

16. 2011, Sparkling Books.

17. http://www.usgovernmentdebt.us offers on the 14th November, 2014 that the Federal Debt of the United States was about $18,006,100,032,000.

18. In his book *Capital in the Twenty-First Century* (Belknap Press, 2014).

19. http://www.marketwatch.com/story/capitalism-is-killing-americas-morals-our-future-2015-05-22.

20. http://en.wikipedia.org/wiki/Population_growth.

21. According to United Nation's 2010 revision to its population

projections, world population will peak at 10.1 billioni n 2100 compared to 7 billion in 2011. A 2014 paper by demographers from several universities and the United Nations Population Division forecast that the world's population will reach about 10.9 billion in 2100 and continue growing thereafter. However, some experts dispute the United Nation's forecast and have argued that birthrates will fall below replacement rate in the 2020s. According to these forecasters, population growth will be only sustained untill the 2040s by rising longevity but will peak below 9 billion by 2050.

22. The Manhattan Project was a research-and-development project that produced the first nuclear weapons during World War II. It was led by the United States with the support of the United Kingdom and Canada. (Wikipedia).

23. http://thebulletin.org/overview.

24. http://en.wikipedia.org/wiki/Doomsday_Clock.

25. Scott Barry Kaufman is scientific director of The Imagination Institute in the Positive Psychology Center at the University of Pennsylvania. Kaufman investigates the development and measurement of imagination, and the many paths to greatness.

26. https://www.psychologytoday.com.

27. Published by Farrar, Straus, and Giroux, 2015

28. http://www.theguardian.com/books/2015/jul/17/postcapitalism-end-of-capitalism-begun

29. Eckhart Tolle, *The Power of Now* (Namaste Publishing, 2004).

30. http://www.theautomaticearth.com/2015/08/power-and-compassion/.

31. https://patriceayme.wordpress.com.

32. *Roget's II The New Thesaurus*, (Houghton Mifflin Company 1995).

33. http://learningfromdogs.com/2011/01/06/just-pause-and/.

34. *Webster's Dictionary*.

35. DR showed me an unaltered photograph taken in 2006 showing Tim lying back on a blanket with his dog across his waist and sitting on its haunches just behind Tim and the dog was Luna, the wolf.

36. Figures from research undertaken in May 2015.

37. http://www.erikdkennedy.com/essays/social-lives-cavemen.php.

38. Erik Kennedy, who gave me permission to publish these extracts, confirmed that the remarkable facts came from Malcolm Gladwell's book *Outliers: The Story of Success* (Back Bay Books, 2011).

39. http://www.washingtonpost.com/national/health-science/in-dogs-play-researchers-see-honesty-and-deceit-perhaps-something-like-morality/2014/05/19/d8367214-ccb3-11e3-95f7-7ecdde72d2ea_story.html.

40. http://davidhgrimm.com.

41. Published by Public Affairs, a member of the Perseus Books Group.

42. http://www.literati.net/authors/marc-bekoff/.

43. https://www.psychologytoday.com/blog/animal-emotions/201505/butts-and-noses-secrets-and-lessons-dog-parks.

44. June 2015 but dropped to nine with the loss of Lilly in August, 2015.

45. It was a famous line from "Meditation XVII.

46. http://www.flagpole.com/about-us.

47. http://www.flagpole.com/news/news-features/2014/10/08/dogs-and-their-homeless-owners-share-love-if-not-shelter.

48. http://www.chron.com/news/houston-texas/article/What-makes-us-human-Teaching-learning-and-3375389.php.

49. http://buynothingproject.org.

50. http://buynothingproject.org/about/.

51. http://opensource.org.

52. http://www.forbes.com/sites/georgebradt/2014/11/25/why-open-leadership-has-become-essential/.

53. http://lifeasahuman.com/2011/pets/a-dog-honestly/.

54. http://teachingamericanhistory.org/library/document/letter-to-alexander-hamilton-2/.

55. http://www.thenewatlantis.com/publications/the-truth-about-human-nature.

56. As in an end or purpose of life.

57. http://contemplatinghappiness.blogspot.com/p/my-books.html.

58. http://www.thefreedictionary.com/openness.

59. http://www.thefreedictionary.com/open-mindedness.

60. I am indebted to Susan Kennedy's writings for inspiring many of the ideas in this chapter.

61. June 2015.

62. http://www.goodtherapy.org.

63. Solution-focused therapy was developed by Steve De Shazer,

Insoo Kim Berg, and their team at the Brief Family Therapy Family Center in Milwaukee, Wisconsin, United States.

64. http://www.jeandonaldson.com.

65. http://www.ted.com/talks/pico_iyer_the_art_of_stillness/transcript?language=en.

66. https://www.ted.com/talks/pico_iyer_the_art_of_stillness?language=en.

67. https://www.ted.com/speakers/pico_iyer.

68. http://www.ted.com/talks/pico_iyer_the_art_of_stillness/transcript?language=en.

69. http://bigthink.com/think-tank/steven-kotler-flow-states.

70. http://www.amazon.com/Anam-Cara-Book-Celtic-Wisdom/dp/006092943X/ref=sr_1_1

71. Refer to Christina Starmans and Paul Bloom of the Mind and Development Lab at Yale University.

72. This is fictional although all the names and places are real.

73. Sharon Salzberg, *Faith, Trusting Your Own Deepest Experience* (Riverhead Books, 2003).

74. http://dogtime.com/dealing-with-grief-of-loss-aaha.html#.

75. http://psychcentral.com/lib/the-5-stages-of-loss-and-grief/000617.

76.http://www.aahanet.org/blog/petsmatter/post/2014/05/20/929363/Life-after-dog-Support-and-resources-on-pet-loss.aspx.

77. http://thebark.com/content/wolf-who-stayed.

Lightning Source UK Ltd.
Milton Keynes UK
UKOW06f0030190216

268662UK00001B/15/P